Undergraduate Texts in Mathematics

Editors

S. Axler

F. W. Gehring

K. A. Ribet

T0192566

Springer

New York
Berlin
Heidelberg
Hong Kong
London
Milan
Paris
Tokyo

Undergraduate Texts in Mathematics

Abbott: Understanding Analysis.

Anglin: Mathematics: A Concise History and Philosophy.
Readings in Mathematics.

Anglin/Lambek: The Heritage of Thales.
Readings in Mathematics.

Apostol: Introduction to Analytic Number Theory. Second edition.

Armstrong: Basic Topology.

Armstrong: Groups and Symmetry.

Axler: Linear Algebra Done Right. Second edition.

Beardon: Limits: A New Approach to Real Analysis.

Bak/Newman: Complex Analysis. Second edition.

Banchoff/Wermer: Linear Algebra Through Geometry. Second edition.

Berberian: A First Course in Real Analysis.

Bix: Conics and Cubics: A Concrete Introduction to Algebraic Curves.

Brémaud: An Introduction to Probabilistic Modeling.

Bressoud: Factorization and Primality Testing.

Bressoud: Second Year Calculus.
Readings in Mathematics.

Brickman: Mathematical Introduction to Linear Programming and Game Theory.

Browder: Mathematical Analysis: An Introduction.

Buchmann: Introduction to Cryptography.

Buskes/van Rooij: Topological Spaces: From Distance to Neighborhood.

Callahan: The Geometry of Spacetime: An Introduction to Special and General Relativity.

Carter/van Brunt: The Lebesgue–Stieltjes Integral: A Practical Introduction.

Cederberg: A Course in Modern Geometries. Second edition.

Childs: A Concrete Introduction to Higher Algebra. Second edition.

Chung: Elementary Probability Theory with Stochastic Processes. Third edition.

Cox/Little/O'Shea: Ideals, Varieties, and Algorithms. Second edition.

Croom: Basic Concepts of Algebraic Topology.

Curtis: Linear Algebra: An Introductory Approach. Fourth edition.

Devlin: The Joy of Sets: Fundamentals of Contemporary Set Theory. Second edition.

Dixmier: General Topology.

Driver: Why Math?

Ebbinghaus/Flum/Thomas: Mathematical Logic. Second edition.

Edgar: Measure, Topology, and Fractal Geometry.

Elaydi: An Introduction to Difference Equations. Second edition.

Erdős/Surányi: Topics in the Theory of Numbers.

Estep: Practical Analysis in One Variable.

Exner: An Accompaniment to Higher Mathematics.

Exner: Inside Calculus.

Fine/Rosenberger: The Fundamental Theory of Algebra.

Fischer: Intermediate Real Analysis.

Flanigan/Kazdan: Calculus Two: Linear and Nonlinear Functions. Second edition.

Fleming: Functions of Several Variables. Second edition.

Foulds: Combinatorial Optimization for Undergraduates.

Foulds: Optimization Techniques: An Introduction.

Franklin: Methods of Mathematical Economics.

Frazier: An Introduction to Wavelets Through Linear Algebra.

(continued after index)

M.A. Armstrong

Groups and Symmetry

With 54 Illustrations

 Springer

M.A. Armstrong
Department of Mathematical Sciences
University of Durham
Durham DH1 3LE
England

Mathematics Subject Classification (2000): 20-01, 20F65

Library of Congress Cataloging-in-Publication Data
Armstrong, M.A. (Mark Anthony)
    Groups and symmetry / M.A. Armstrong.
      (Undergraduate texts in mathematics)
      p.   cm.
    Bibliography: p.
    Includes index.
    1. Groups, Theory of.   2. Symmetry groups.   I. Title.
QA171.A76   1988
512'.2—dc19   87-37677
  ISBN 978-1-4419-3085-9
Printed on acid-free paper.

Cover art taken from *Ornamental Design* by Claude Humbert © Office du Livre, Freibourg, Switzerland.

Printed in the United States of America.   (ASC/SBV)

9 8 7

Springer-Verlag is a part of *Springer Science+Business Media*

springeronline.com

For Jerome and Emily

The beauty of a snow crystal depends on its mathematical regularity and symmetry; but somehow the association of many variants of a single type, all related but no two the same, vastly increases our pleasure and admiration.

D'ARCY THOMPSON
(*On Growth and Form*, Cambridge, 1917.)

En général je crois que les seules structures mathématiques intéressantes, dotées d'une certaine légitimité, sont celles ayant une réalisation naturelle dans le continu.... Du reste, cela se voit très bien dans des théories purement algébriques comme la théorie des groupes abstraits ou on a des groupes plus ou moins étranges apparaissant comme des groupes d'automorphismes de figures continues.

RENÉ THOM
(*Paraboles et Catastrophes*, Flammarion, 1983.)

Preface

Numbers measure size, *groups measure symmetry*. The first statement comes as no surprise; after all, that is what numbers "are for". The second will be exploited here in an attempt to introduce the vocabulary and some of the highlights of elementary group theory.

A word about content and style seems appropriate. In this volume, the emphasis is on *examples* throughout, with a weighting towards the symmetry groups of solids and patterns. Almost all the topics have been chosen so as to show groups in their most natural role, acting on (or permuting) the members of a set, whether it be the diagonals of a cube, the edges of a tree, or even some collection of subgroups of the given group. The material is divided into twenty-eight short chapters, each of which introduces a new result or idea. A glance at the Contents will show that most of the mainstays of a "first course" are here. The theorems of Lagrange, Cauchy, and Sylow all have a chapter to themselves, as do the classification of finitely generated abelian groups, the enumeration of the finite rotation groups and the plane crystallographic groups, and the Nielsen–Schreier theorem.

I have tried to be informal wherever possible, listing only significant results as theorems and avoiding endless lists of definitions. My aim has been to write a book which can be read with or without the support of a course of lectures. It is not designed for use as a dictionary or handbook, though new concepts are shown in bold type and are easily found in the index. Every chapter ends with a collection of exercises designed to consolidate, and in some cases fill out, the main text. It is essential to work through as many of these as possible before moving from one chapter to the next. Mathematics is not for spectators; to gain in understanding, confidence, and enthusiasm one has to participate.

As prerequisites I assume a first course in linear algebra (including matrix multiplication and the representation of linear maps between Euclidean spaces by matrices, though not the abstract theory of vector spaces) plus familiarity with the basic properties of the real and complex numbers. It would seem a pity to teach group theory without matrix groups available as a rich source of examples, especially since matrices are so heavily used in applications.

Elementary material of this type is all common stock, nevertheless it is not static, and improvements are made from time to time. Three such should be mentioned here: H. Wielandt's approach to the Sylow theorems (Chapter 20), James H. McKay's proof of Cauchy's theorem (Chapter 13), and the introduction of groups acting on trees by J.-P. Serre (Chapter 28). Another influence is of a more personal nature. As a student I had the good fortune to study with A.M. Macbeath, whose lectures first introduced me to group theory. The debt of gratitude from pupil to teacher is best paid in kind. If this little book can pass on something of the same appreciation of the beauty of mathematics as was shown to me, then I shall be more than satisfied.

Durham, England   M.A.A.
September 1987

Acknowledgements

My thanks go to Andrew Jobbings who read and commented on much of the manuscript, to Lyndon Woodward for many stimulating discussions over the years, to Mrs. S. Nesbitt for her good humour and patience whilst typing, and to the following publishers who have kindly permitted the use of previously published material: Cambridge University Press (quotation from *On Growth and Form*), Flammarion (quotation from *Paraboles et Catastrophes*), Dover Publications (Figure 2.1 taken from *Snow Crystals*), Office du Livre, Fribourg (Figure 25.3 and parts of Figure 26.2 taken from *Ornamental Design*), and Plenum Publishing Corporation (parts of Figure 26.2 taken from *Symmetry in Science and Art*).

Contents

CHAPTER 1

Symmetries of the Tetrahedron

How much symmetry has a tetrahedron? Consider a regular tetrahedron T and, for simplicity, think only of rotational symmetry. Figure 1.1 shows two axes. One, labelled L, passes through a vertex of the tetrahedron and through the centroid of the opposite face; the other, labelled M, is determined by the midpoints of a pair of opposite edges. There are four axes like L and two rotations about each of these, through $2\pi/3$ and $4\pi/3$, which send the tetrahedron to itself. The sense of the rotations is as shown: looking along the axis from the vertex in question the opposite face is rotated anticlockwise. Of course, rotating through $2\pi/3$ (or $4\pi/3$) in the opposite sense has the same effect on T as our rotation through $4\pi/3$ (respectively $2\pi/3$). As for axis M, all we can do is rotate through π, and there are three axes of this kind. So far we have $(4 \times 2) + 3 = 11$ symmetries. Throwing in the identity symmetry, which leaves T fixed and is equivalent to a full rotation through 2π about any of our axes, gives a total of twelve rotations.

We seem to have answered our original question. There are precisely twelve rotations, counting the identity, which move the tetrahedron onto itself. But this is not the end of the story. A flat hexagonal plate with equal sides also has twelve rotational symmetries (Fig. 1.2), as does a right regular pyramid on a twelve sided base (Fig. 1.3). For the plate we have five rotations (through $\pi/3$, $2\pi/3$, π, $4\pi/3$, and $5\pi/3$) about the axis perpendicular to it which passes through its centre of gravity. In addition there are three axes of symmetry determined by pairs of opposite corners, three determined by the midpoints of pairs of opposite sides, and we can rotate the plate through π about each of these. Not forgetting the identity, our total is again twelve. The pyramid has only one axis of rotational symmetry. It joins the apex of the pyramid to the centroid of its base, and there are twelve distinct rotations (through $k\pi/6$, $1 \leqslant k \leqslant 12$, in some chosen sense) about this axis. Despite the fact that we

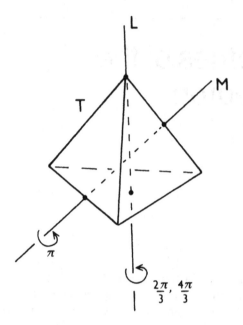

Figure 1.1

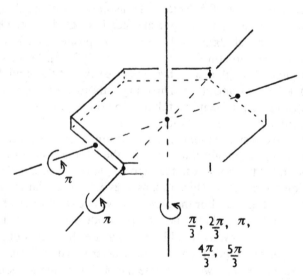

Figure 1.2

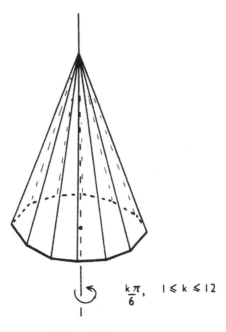

$$\frac{k\pi}{6}, \quad 1 \leqslant k \leqslant 12$$

Figure 1.3

have counted twelve rotations in each case, the tetrahedron, the plate, and the pyramid quite clearly do not exhibit the same symmetry.

The most striking difference is that the pyramid possesses just one axis of symmetry. A rotation of $\pi/6$ about this axis has to be repeated (in other words, combined with itself) twelve times before the pyramid returns to its original position. Indeed, by suitable repetition of this basic rotation we can produce all the other eleven symmetries. However, no single rotation of the plate or the tetrahedron when repeated will give us all the other rotations.

If we look more carefully we can spot other differences, all of which have to do, in one way or another, with the way in which our symmetries combine. For example, the symmetries of the pyramid all *commute* with each other. That is to say, if we take any two and perform one rotation after the other, the effect on the pyramid is the same no matter which one we choose to do first. (These rotations all have the same axis, so if, for the sake of argument, we rotate through $\pi/3$ then through $5\pi/6$, we obtain rotation through $7\pi/6$, which is also the result of $5\pi/6$ first followed by $\pi/3$.) This is not the case for the tetrahedron or the plate. We recommend an experiment with the tetrahedron. Labelling the vertices of T as in Figure 1.4 enables us to see clearly the effect of a particular symmetry. Think of the rotations r ($2\pi/3$ about axis L in the sense indicated) and s (π about axis M). Performing first r then s takes vertex 2 back to its initial position and gives a rotation about axis N. But first s then r moves 2 to the place originally occupied by 4, and so cannot be the same rotation. Do

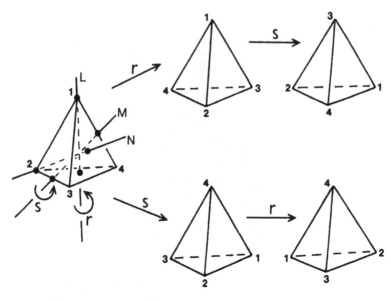

Figure 1.4

not fall into the trap of carrying the axis of s along with you as you do r first. Both r and s should be thought of as rigid motions of space, each of which has an axis that is *fixed* in space, and each of which rotates T onto itself.

Here is a third observation. There is only one rotation of the pyramid which, when combined once with itself, gives the identity; namely, the unique rotation through π. The plate has seven such symmetries and the tetrahedron three. These three rotations through π of the tetrahedron commute with one another, but only one of the seven belonging to the plate commutes with all the other six. Which one? Experiment until you find out.

To obtain a decent measure of symmetry, simply counting symmetries is not enough; we must also take into consideration how they combine with each other. It is the so-called symmetry group which captures this information and which we now attempt to describe.

The set of rotational symmetries of T has a certain amount of "algebraic structure". Given two rotations u and v we can *combine* them, by first doing v, then doing u, to produce a new rotation which also takes T to itself, and which we write uv. (Our choice of uv rather than vu is influenced by the convention for the composition of two functions, where fg usually means first apply g, then apply f.) The *identity* rotation, which we denote by e, behaves in a rather special way. Applying first e then another rotation u, or first u then e, always gives the same result as just applying u. In other words $ue = u$ and $eu = u$ for every symmetry u of T. Each rotation u has a so-called *inverse* u^{-1}, which is also a symmetry of T and which satisfies $u^{-1}u = e$ and $uu^{-1} = e$. To obtain u^{-1}, just rotate about the same axis and through the same angle as for u, but in the opposite sense. (For example, the inverse of the rotation r is rr, because applying r three times gives the identity.) Finally, if we take three of our rotations u, v, and w, it does not matter whether we first do w then the composite rotation uv, or whether we apply vw first and then u. In symbols this reduces to $(uv)w = u(vw)$ for any three (not necessarily distinct) symmetries of T.

The twelve symmetries of the tetrahedron together with this algebraic structure make up its rotational symmetry group.

EXERCISES

1.1. Glue two copies of a regular tetrahedron together so that they have a triangular face in common, and work out all the rotational symmetries of this new solid.

1.2. Find all the rotational symmetries of a cube.

1.3. Adopt the notation of Figure 1.4. Show that the axis of the composite rotation srs passes through vertex 4, and that the axis of $rsrr$ is determined by the midpoints of edges 12 and 34.

1.4. Having completed the previous exercise, express each of the twelve rotational symmetries of the tetrahedron in terms of r and s.

1.5. Again with the notation of Figure 1.4, check that $r^{-1} = rr$, $s^{-1} = s$, $(rs)^{-1} = srr$, and $(sr)^{-1} = rrs$.

1.6. Show that a regular tetrahedron has a total of twenty-four symmetries if reflections and products of reflections are allowed. Identify a symmetry which is not a rotation and not a reflection. Check that this symmetry is a product of three reflections.

1.7. Let q denote reflection of a regular tetrahedron in the plane determined by its centroid and one of its edges. Show that the rotational symmetries, together with those of the form uq, where u is a rotation, give all twenty-four symmetries of the tetrahedron.

1.8. Find all plane symmetries (rotations and reflections) of a regular pentagon and of a regular hexagon.

1.9. Show that the hexagonal plate of Figure 1.2 has twenty-four symmetries in all. Identify those symmetries which commute with all the others.

1.10. Make models of the octahedron, dodecahedron, and icosahedron (see Fig. 8.1). Try to spot as many symmetries of each of these solids as you can.

CHAPTER 2

Axioms

Without further ado we define the notion of a group, using the symmetries of the tetrahedron as guide. The first ingredient is a set. The second is a rule which allows us to combine any ordered pair x, y of elements from the set and obtain a unique "product" xy *which also lies in the set.* This rule is usually referred to as a "multiplication" on the given set.

A **group** *is a set G together with a multiplication on G which satisfies three axioms:*

(a) *The multiplication is associative, that is to say* $(xy)z = x(yz)$ *for any three (not necessarily distinct) elements from G.*

(b) *There is an element e in G, called an identity element, such that* $xe = x = ex$ *for every x in G.*

(c) *Each element x of G has a (so-called) inverse* x^{-1} *which belongs to the set G and satisfies* $x^{-1}x = e = xx^{-1}$.

How does a formal definition couched in terms of axioms help? So far not at all; indeed, if the only group turned out to be the rotational symmetry group of the tetrahedron, we would be wasting our time. But this is not the case; groups crop up in many different situations.

All of us take the additive group structure of the set of real numbers for granted. Here the rule for combining an ordered pair of numbers x, y is simply to add them to give $x + y$. We accept that $(x + y) + z = x + (y + z)$ for any three real numbers, there is an identity element, namely, zero, and $- x$ is clearly an inverse for the real number x. This example shows why we previously placed the words product and multiplication in quotation marks. The rule which enables us to combine our elements is invariably referred to as a multiplication, but may have nothing to do with multiplication of numbers in the usual sense.

A chemist may be interested in the amount of symmetry possessed by a particular molecule. Methane (CH_4), for example, can be thought of as having a carbon nucleus at the centroid of a regular tetrahedron, with four protons (hydrogen nuclei) arranged at the vertices. The benzene molecule (C_6H_6), on the other hand, is modelled by a hexagonal structure with a carbon and a hydrogen nucleus at each vertex. (Hexagonal symmetry is common in nature, perhaps nowhere more pleasing than in the structure of a snow crystal; see Fig. 2.1.) From our experience with the tetrahedron and the hexagon we know that it matters in which order we combine two symmetries. Hence, the continual reference to *ordered* pairs of elements. It matters whether we take two elements of a group in the order x, y or in the opposite order y, x. In the first case our rule gives the answer xy, in the second yx, and these two need not be equal.

A physicist learning relativity meets the Lorentz group, whose elements are matrices of the form

$$\begin{bmatrix} \cosh u & \sinh u \\ \sinh u & \cosh u \end{bmatrix} \qquad (*)$$

and which are combined via matrix multiplication. Remember that cosh u, sinh u are the hyperbolic functions, so called because the equations $x = \cosh u$, $y = \sinh u$ determine the hyperbola $x^2 - y^2 = 1$. They satisfy

$$\cosh(u \pm v) = \cosh u \cosh v \pm \sinh u \sinh v,$$

$$\sinh(u \pm v) = \sinh u \cosh v \pm \cosh u \sinh v$$

consequently,

$$\begin{bmatrix} \cosh u & \sinh u \\ \sinh u & \cosh u \end{bmatrix} \begin{bmatrix} \cosh v & \sinh v \\ \sinh v & \cosh v \end{bmatrix} = \begin{bmatrix} \cosh(u+v) & \sinh(u+v) \\ \sinh(u+v) & \cosh(u+v) \end{bmatrix}$$

and this product does give a matrix *of the same form*. The identity matrix fulfils the requirements of an identity, and lies in the given set of matrices because it is equal to

$$\begin{bmatrix} \cosh 0 & \sinh 0 \\ \sinh 0 & \cosh 0 \end{bmatrix}$$

As an inverse for $(*)$ we can use

$$\begin{bmatrix} \cosh(-u) & \sinh(-u) \\ \sinh(-u) & \cosh(-u) \end{bmatrix}$$

which has the required form. Since matrix multiplication is associative, we have a group.

A mathematician thinking about Euclidean geometry finds he is studying those properties of figures which are left unchanged by the elements of a

Figure 2.1

particular group, the group of *similarities* of the plane. A similarity enlarges or shrinks figures while keeping them the same shape. More precisely, it sends straight-line segments to straight-line segments, multiplying their lengths by a factor which is the same for every segment. Triangles are sent to similar triangles, angles being preserved in magnitude, though not necessarily in sense. The composition of two similarities is another, and the group axioms are easily checked (see Exercise 2.4).

It is precisely when we recognise the same amount of structure in a wide variety of interesting examples that the abstract approach comes into its own. Starting from the axioms for a group, we shall build up a body of results which may be used whenever these axioms are satisfied, a much more satisfactory state of affairs than having to verify a specific property time and time again for different groups.

Here are two properties common to all groups. *The identity element of a group is unique.* Suppose two elements e and e' are both identities. Then $ee' = e'$ because e is an identity, and $ee' = e$ because e' is an identity. Hence e is equal to e'. *The inverse of each element of a group is unique.* Assume y and z are both inverses for the element x. Then

$$y = ey \qquad \text{(where } e \text{ is the identity in the group)}$$

$$= (zx)y \qquad \text{(since } z \text{ is an inverse for } x\text{)}$$

$$= z(xy) \qquad \text{(because the multiplication is associative)}$$

$$= ze \qquad \text{(since } y \text{ is also an inverse for } x\text{)}$$

$$= z \qquad \text{(as } e \text{ is the identity).}$$

Hence y is equal to z, and the inverse of x is indeed unique. Notice that both arguments use only those facts about a group which are supplied by the axioms. For this reason we can be confident that the conclusions hold for every group.

In the next few sections we shall begin to develop theoretical results alongside concrete examples of groups. Remember, *the examples are important*; without them the theory is at best a poor form of intellectual entertainment.

EXERCISES

2.1. Compare the symmetry of a snow crystal with that of the hexagonal plate in Figure 1.2.

2.2. Show that the set of positive real numbers forms a group under multiplication.

2.3. Which of the following collections of 2 × 2 matrices with real entries form groups under matrix multiplication?

(i) Those of the form $\begin{bmatrix} a & b \\ b & c \end{bmatrix}$ for which $ac \neq b^2$.

(ii) Those with entries $\begin{bmatrix} a & b \\ c & a \end{bmatrix}$ such that $a^2 \neq bc$.

(iii) Those of the form $\begin{bmatrix} a & b \\ 0 & c \end{bmatrix}$ where ac is not zero.

(iv) Those which have non-zero determinant and whose entries are integers.

2.4. Let f be a similarity of the plane. Show that f is a bijection and that the inverse function f^{-1} is also a similarity. Verify that the collection of all similarities of the plane forms a group under composition of functions.

2.5. A function from the plane to itself which preserves the distance between any two points is called an *isometry*. Prove that an isometry must be a bijection and check that the collection of all isometries of the plane forms a group under composition of functions.

2.6. Show that the collection of all rotations of the plane about a fixed point P forms a group under composition of functions. Is the same true of the set of all reflections in lines which pass through P? What happens if we take all the rotations and all the reflections?

2.7. Let x and y be elements of a group G. Prove that G contains elements w, z which satisfy $wx = y$ and $xz = y$, and show that these elements are unique.

2.8. If x and y are elements of a group, prove that $(xy)^{-1} = y^{-1}x^{-1}$.

CHAPTER 3

Numbers

Perhaps the quickest way to get used to the group axioms is to look at some groups of numbers. The list below serves to give examples and to establish some notation.

Addition of numbers (real or complex) makes each of the following sets into a group:

$\mathbb{Z}$, the set of integers;
$\mathbb{Q}$, the set of rational numbers;
$\mathbb{R}$, the set of real numbers;
$\mathbb{C}$, the set of complex numbers.

In each case zero is the identity element, and $-x$ is the inverse of the number x.

Multiplication of numbers (real or complex) makes each of the following sets into a group:

$\mathbb{Q} - \{0\}$, the non-zero rationals;
$\mathbb{R} - \{0\}$, the non-zero reals;
$\mathbb{Q}^{\text{pos}}$, the positive rationals;
$\mathbb{R}^{\text{pos}}$, the positive reals;
$\{+1, -1\}$;
$\mathbb{C} - \{0\}$, the non-zero complex numbers;
C, the complex numbers of modulus 1;
$\{\pm 1, \pm i\}$.

In each case the number 1 is the identity element, and $1/x$ is the inverse of the number x.

It is worth examining this list in some detail, as much for what is missing as for what it contains. Adding two integers always produces an integer. This is the first thing we need to notice when checking that the integers form a group under addition. But if we took, say, the set of all odd integers, and again used addition as the rule for combining elements, the result could not be a group because the sum of two odd integers is even and so does not belong to the given set.

Turning to multiplication of numbers as group multiplication, we must remove zero from the set of real numbers if we wish to have a group. There is clearly no number x such that $x \cdot 0 = 1$; in other words, zero does not have a multiplicative inverse. It is easy to check that the non-zero real, rational, and complex numbers all form groups under multiplication. What about the non-zero integers? Multiplication does not make them into a group. The only number x which satisfies $2 \cdot x = 1$ is $\frac{1}{2}$, which is not an integer. Therefore, 2 has no multiplicative inverse in the set of integers.

We use C to denote the unit circle in the complex plane, that is to say, the set of those complex numbers which have modulus one. If $z, w \in C$, then $|zw| = |z||w| = 1$, showing that $zw \in C$. The number 1 lies in C and acts as identity for complex multiplication. Finally, if $z \in C$, then $|1/z| = 1/|z| = 1$, so $1/z \in C$ and each element of C has a multiplicative inverse which also belongs to C. Therefore, complex multiplication makes C into a group. We have made no mention of the associative law, but if we accept that it holds for multiplication of any three complex numbers, then it certainly holds for any three numbers taken from C.

Strictly speaking, we should use notation such as $(\mathbb{R}, +)$ to denote the additive group of reals, making it absolutely clear that the underlying set is the set of real numbers, and the group "multiplication" is addition of numbers. In practice, this is cumbersome to work with, so we agree to use the same symbol $\mathbb{R}$ for the set of real numbers, and for the group of real numbers under addition. It will usually be clear from the context which one we mean. The other symbols introduced in our list will also be used to stand for the corresponding groups.

The set of integers is a subset of the set of real numbers, and both form groups under addition. We shall say that $\mathbb{Z}$ is a "subgroup" of $\mathbb{R}$. This idea will be taken up again in Chapter 5.

A group is *commutative*, or **abelian**, if $xy = yx$ for any two of its elements. All of the examples in our list are abelian because $x + y = y + x$ and $x \cdot y = y \cdot x$ for any two numbers x, y, real or complex.

Let n be a positive integer. The set $0, 1, 2, \ldots, n-1$ can be made into a group using *addition modulo n*. That is to say, if x and y are members of this set, define

$$x +_n y = \begin{cases} x + y & \text{if } 0 \leqslant x + y < n \\ x + y - n & \text{if } x + y \geqslant n \end{cases}$$

and use this as "multiplication". For example, $5 +_6 3 = 8 - 6 = 2$. (Counting modulo a particular number is a familiar idea; think of adding angles modulo 2π.) The group axioms are easy to check. This sum $x +_n y$ is again an integer between 0 and $n - 1$. Both $(x +_n y) +_n z$ and $x +_n (y +_n z)$ are equal to

$$x + y + z \qquad \text{if } 0 \leqslant x + y + z < n$$

$$x + y + z - n \qquad \text{if } n \leqslant x + y + z < 2n$$

$$x + y + z - 2n \qquad \text{if } x + y + z \geqslant 2n$$

so associativity follows. Zero is the identity element, and $n - x$ is the inverse of x for $x \neq 0$. The group is abelian because $x +_n y = y +_n x$. Therefore we have a finite abelian group with n elements which will be denoted by $\mathbb{Z}_n$.

Two integers are *congruent modulo n* if they differ by a multiple of n. Of course each integer x is congruent modulo n to exactly one of the integers 0, $1, 2, \ldots, n - 1$, namely, to the remainder obtained on dividing x by n. We shall refer to this remainder as "x read mod n", or simply $x (\text{mod } n)$. Then $x +_n y$ is just $x + y$ read mod n.

The integers $0, 1, 2, \ldots, n - 1$ may also be multiplied modulo n via

$$x \cdot_n y = xy (\text{mod } n).$$

For example $5 \cdot_6 3 = 3$ because dividing fifteen by six leaves remainder three. Can we obtain a group using this multiplication? As usual we must remove the number zero. But this may not be enough. Take n to be ten. Then $2 \cdot_{10} 5 = 0$, so multiplication modulo ten of two numbers from $1, 2, \ldots, 9$ does not always produce another number between 1 and 9. Therefore, we do not have a group. In fact, multiplication modulo n makes the integers $1, 2, \ldots, n - 1$ into a group precisely when n is a *prime* number (Exercise 3.10). A simple experiment shows that deleting the integers 0, 2, 4, 5, 6, 8 leaves a collection which do have the structure of a group when multiplied modulo ten. Does this suggest a general result? (The answer can be found in Chapter 11.)

EXERCISES

3.1. Show that each of the following collections of numbers forms a group under addition.

(i) The even integers.
(ii) All real numbers of the form $a + b\sqrt{2}$ where $a, b \in \mathbb{Z}$.
(iii) All real numbers of the form $a + b\sqrt{2}$ where $a, b \in \mathbb{Q}$.
(iv) All complex numbers of the form $a + bi$ where $a, b \in \mathbb{Z}$.

3.2. Write $\mathbb{Q}(\sqrt{2})$ for the set described in Exercise 3.1 (iii). Given a non-zero element $a + b\sqrt{2}$ of $\mathbb{Q}(\sqrt{2})$, express $1/(a + b\sqrt{2})$ in the form $c + d\sqrt{2}$, where $c, d \in \mathbb{Q}$. Prove that multiplication makes $\mathbb{Q}(\sqrt{2}) - \{0\}$ into a group.

3.3. Let n be a positive integer and let G consist of all those complex numbers z which satisfy $z^n = 1$. Show that G forms a group under multiplication of complex numbers.

3.4. Vary n in the previous exercise and check that the union of all these groups

$$\bigcup_{n=1}^{\infty} \{z \in \mathbb{C} \mid z^n = 1\}$$

is also a group under multiplication of complex numbers.

3.5. Let n be a positive integer. Prove that

$$(x \cdot_n y) \cdot_n z = x \cdot_n (y \cdot_n z)$$

for all $x, y, z \in \mathbb{Z}$.

3.6. Verify that each of the sets

$$\{1, 3, 7, 9, 11, 13, 17, 19\}$$
$$\{1, 3, 7, 9\}$$
$$\{1, 9, 13, 17\}$$

forms a group under multiplication modulo 20.

3.7. Which of the following sets form groups under multiplication modulo 14?

$$\{1, 3, 5\}, \qquad \{1, 3, 5, 7\}$$
$$\{1, 7, 13\}, \qquad \{1, 9, 11, 13\}.$$

3.8. Show that if a subset of $\{1, 2, \ldots, 21\}$ contains an even number, or contains the number 11, then it cannot form a group under multiplication modulo 22.

3.9. Let p be a prime number and let x be an integer which satisfies $1 \leqslant x \leqslant p - 1$. Show that none of $x, 2x, \ldots, (p-1)x$ is a multiple of p. Deduce the existence of an integer z such that $1 \leqslant z \leqslant p - 1$ and $xz \pmod p = 1$.

3.10. Use the results of Exercises 3.5 and 3.9 to verify that multiplication modulo n makes $\{1, 2, \ldots, n-1\}$ into a group if n is prime. What goes wrong when n is not a prime number?

CHAPTER 4

Dihedral Groups

Think back to the flat hexagonal plate mentioned earlier. Its twelve rotational symmetries combine in the natural way to form a group. For each positive integer n greater than or equal to three we can manufacture a plate which has n equal sides. In this way we produce a family of symmetry groups which are not commutative, the so-called **dihedral groups**.

When n is three we have a triangular plate. It has six rotational symmetries and, if r and s are the rotations shown in Figure 4.1, they are

$$e, r, r^2, s, rs, r^2s. \qquad (*)$$

Here r^2 is shorthand for rr and means carry out r twice. Clearly r^3 is the identity, since repeating r three times gives a full rotation through 2π, and higher powers of r will not give anything new. Of course, s^2 is also the identity. Remember our convention that rs means the symmetry obtained by first applying s to the triangle, then applying r. If we do this, as in Figure 4.2, we see that rs is rotation through π about the axis of symmetry labelled M. Similarly r^2s is rotation through π about axis N.

The six elements $(*)$ form a group denoted by D_3. So if we take two of them and combine them in either order, we should in each case obtain a member of the group. We see rs listed, but where is sr? A glance at Figure 4.2 again shows it is equal to r^2s. Similarly $sr^2 = rs$: this may be checked geometrically, or algebraically as follows:

$$sr^2 = s(rr) = (sr)r = (r^2s)r = r^2(sr)$$
$$= r^2(r^2s) = r^4s = r^3(rs) = e(rs)$$
$$= rs.$$

Notice the repeated use of the associative law. We have made quite a meal of

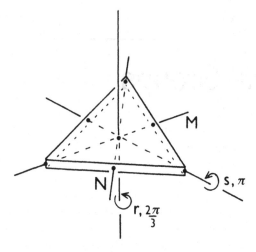

Figure 4.1

this calculation, with a little practice one would not include all the steps! Our aim is to illustrate that knowing $r^3 = e$, $s^2 = e$, and $sr = r^2s$ allows us to manipulate any product and produce one of the six elements of the list (∗). Here are two more examples:

$$r(r^2s) = r^3s = es = s;$$

$$(r^2s)(rs) = r^2(s(rs)) = r^2((sr)s) = r^2((r^2s)s)$$
$$= r^2(r^2s^2) = r^4s^2 = re = r.$$

The first step in the previous calculation could have been $(r^2s)(rs) = ((r^2s)r)s$. Hopefully the answer would be the same. In fact, a product such as r^2srs is independent of the way in which we choose to bracket its terms. More

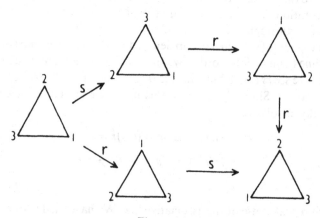

Figure 4.2

generally, *if x_1, x_2, ..., x_n are elements of a group, any two ways of combining these elements in this order give the same answer*. In other words the product $x_1 x_2 ... x_n$ makes sense without any brackets. A proof by induction is outlined in Exercise 4.10. The inductive step uses the associative law.

Calculations can now be carried out with less fuss, for example

$$xyy^{-1}x^{-1} = xex^{-1} = xx^{-1} = e,$$

and similarly $y^{-1}x^{-1}xy = e$. Therefore, *if x and y are elements of a group, then* $(xy)^{-1} = y^{-1}x^{-1}$. In the same way, *if x_1, x_2, ..., x_n are elements of a group, then*

$$(x_1 x_2 ... x_n)^{-1} = x_n^{-1} ... x_2^{-1} x_1^{-1}$$

Write x^m for the product of m copies of the element x, and x^{-n} for the product of n copies of x^{-1}. Then $x^m x^n = x^{m+n}$ and $(x^m)^n = x^{mn}$ for any two integers m and n, provided we interpret x^0 as the identity.

The following table, called a *multiplication table*, shows all 36 possible products xy of ordered pairs of elements x, y taken from D_3

	e	r	r^2	s	rs	r^2s
e	e	r	r^2	s	rs	r^2s
r	r	r^2	e	rs	r^2s	s
r^2	r^2	e	r	r^2s	s	rs
s	s	r^2s	rs	e	r^2	r
rs	rs	s	r^2s	r	e	r^2
r^2s	r^2s	rs	s	r^2	r	e

The product xy lies at the intersection of row x with column y. For example the entry circled is $s(rs)$. Notice each element of the group appears exactly once in every row and every column of the table. This is true of the multiplication table of any group (see Exercise 4.4). In particular, the identity occurs exactly once in each row, corresponding to the fact that each element of a group has a unique inverse.

The dihedral group D_n is the rotational symmetry group of a flat plate with n equal sides. Its elements can be described in the same manner as that used for D_3. Let r be a rotation of the plate through $2\pi/n$ about the axis of symmetry perpendicular to the plate, and s rotation through π about an axis of symmetry which lies in the plane of the plate. Then the elements of D_n are

$$e, r, r^2, ..., r^{n-1}, s, rs, r^2s, ..., r^{n-1}s.$$

Clearly $r^n = e$, $s^2 = e$, and we can check geometrically that $sr = r^{n-1}s$. Since $r^{n-1} = r^{-1}$, we usually write this last relation as $sr = r^{-1}s$. As before, all other products can be worked out using these. For example,

$$sr^2 = srr = r^{-1}sr = r^{-2}s = r^{n-2}s.$$

Each element of the group has the form r^a or $r^a s$ where $0 \leqslant a \leqslant n - 1$ and we find that

$$\left. \begin{array}{l} r^a r^b = r^k \\ r^a(r^b s) = r^k s \end{array} \right\} \quad \text{where } k = a +_n b,$$

$$\left. \begin{array}{l} (r^a s)r^b = r^l s \\ (r^a s)(r^b s) = r^l \end{array} \right\} \quad \text{where } l = a +_n (n - b).$$

We say that r and s together "generate" the group D_n, an idea that will be developed in Chapter 5.

The **order** of a finite group is the number of elements in the group. A group that contains infinitely many elements is said to have *infinite order*. We usually write $|G|$ for the order of the group G. If x is an element of a group, and if $x^n = e$ for some positive integer n, then we say x has finite order, and the smallest positive integer m such that $x^m = e$ is called the *order of* x. Otherwise x has infinite order.

EXAMPLES.

(i) The order of D_3 is six. There are two elements of order three (r, r^2) and three elements of order two $(s, rs, r^2 s)$.

(ii) The order of $\mathbb{Z}_6$ is also six. The elements 1 and 5 both have order six, 2 and 4 have order three, and 3 has order two.

(iii) $\mathbb{R}$ has infinite order, and every element (except 0) has infinite order because repeatedly adding a real number to itself never gives zero, unless of course the number was zero to start with.

(iv) C is the unit circle in the complex plane made into a group by multiplication of complex numbers. It is an infinite group and has elements of both finite and infinite order. A typical element $e^{i\theta}$ has finite order precisely when θ is a rational multiple of 2π, that is to say when $\theta = 2m\pi/n$ for some integers m and n.

EXERCISES

4.1. Work out the multiplication table of the dihedral group D_4. How many elements of order 2 are there in D_n?

4.2. Find the order of each element of $\mathbb{Z}_5$, $\mathbb{Z}_9$, and $\mathbb{Z}_{12}$.

4.3. Check that the integers 1, 2, 4, 7, 8, 11, 13, 14 form a group under multiplication modulo 15. Work out its multiplication table and find the order of each element.

4.4. Let g be an element of a group G. Keep g fixed and let x vary through G. Prove that the products gx are all distinct and fill out G. Do the same for the products xg.

4.5. An element x of a group satisfies $x^2 = e$ precisely when $x = x^{-1}$. Use this observation to show that a group of even order must contain an odd number of elements of order 2.

4.6. If x, y are elements of a group G, and if all three of x, y, xy have order 2, prove that $xy = yx$.

4.7. Let G be the collection of all rational numbers x which satisfy $0 \leqslant x < 1$. Show that the operation

$$x + y = \begin{cases} x + y & \text{if } 0 \leqslant x + y < 1 \\ x + y - 1 & \text{if } x + y \geqslant 1 \end{cases}$$

makes G into an infinite abelian group all of whose elements have finite order.

4.8. Let x and g be elements of a group G. Show that x and gxg^{-1} have the same order. Now prove that xy and yx have the same order for any two elements x, y of G.

4.9. Check that the 2×2 matrices

$$\begin{bmatrix} a & b \\ c & d \end{bmatrix} \quad \text{for which } a, b, c, d \in \mathbb{Z} \text{ and } ad - bc = 1$$

form a group under matrix multiplication. Let

$$A = \begin{bmatrix} 0 & -1 \\ 1 & 0 \end{bmatrix}, \quad B = \begin{bmatrix} 0 & 1 \\ -1 & -1 \end{bmatrix}$$

and find the orders of A, B, AB, BA.

4.10. *General associative law.* Let G be a group and assume inductively that products $x_1 x_2 \ldots x_k$ of elements of G always make sense without any brackets provided $1 \leqslant k \leqslant n - 1$. We must verify that an arbitrary product $x_1 x_2 \ldots x_n$ of length n is well defined regardless of the way in which we bracket its terms. Suppose we combine these elements in two different ways, and that the final multiplications in the two procedures are

$$(x_1 x_2 \ldots x_r)(x_{r+1} \ldots x_n), \tag{1}$$

$$(x_1 x_2 \ldots x_s)(x_{s+1} \ldots x_n), \tag{2}$$

where $1 \leqslant r < s \leqslant n - 1$. These terms inside brackets make sense by our inductive hypothesis. Write (1) as

$$(x_1 x_2 \ldots x_r)[(x_{r+1} \ldots x_s)(x_{s+1} \ldots x_n)]$$

express (2) in a similar fashion, and use the ordinary associative law for three elements to finish the argument.

CHAPTER 5

Subgroups and Generators

Inside D_6 the six elements

$$e, r^2, r^4, s, r^2s, r^4s$$

form a group with respect to composition of symmetries. This is easy to check. The product of any two of these gives another, the identity is present, and since

$$e^{-1} = e, (r^2)^{-1} = r^4, (r^4)^{-1} = r^2, s^{-1} = s, (r^2s)^{-1} = r^2s, (r^4s)^{-1} = r^4s$$

all the inverses are also present. If we look at Figure 5.1 we see that these elements form the rotational symmetry group of a triangle inscribed inside the hexagon. So they make up a "copy" of D_3 sitting inside D_6, a so called subgroup of D_6 in the following sense.

*A **subgroup** of a group G is a subset of G which itself forms a group under the multiplication of G.*

(When we use the symbol G to denote a group we must remember that G carries with it a "multiplication" and is not just a set of elements.)

Suppose we have a subset H of G, and want to know whether or not it is a subgroup. Then we must ask three questions. Given two elements x, y of H we can form their product xy in G. Does this product always belong to H? Does the identity element of G belong to H? Each element of H certainly has an inverse in G; does this inverse always belong to H? If the answer to all three questions is yes, then H forms a group with respect to the multiplication of G, and is therefore a subgroup of G. Notice: It is not necessary to check the associative law. For if $(xy)z$ and $x(yz)$ are equal for any three elements of G, they are certainly equal for any three elements chosen from a subset of G. When H is a subgroup of G we write $H < G$.

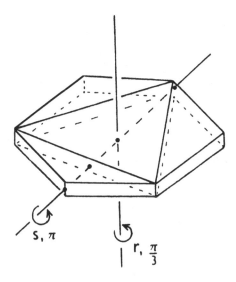

Figure 5.1

EXAMPLES.

(i) $\mathbb{Z} < \mathbb{Q}$, $\mathbb{Q} < \mathbb{R}$, and $\mathbb{R} < \mathbb{C}$.

(ii) The even integers, which we denote by $2\mathbb{Z}$, form a subgroup of the additive group of integers. For any positive integer n, the set $n\mathbb{Z}$ of all integer multiples of n is a subgroup of $\mathbb{Z}$.

(iii) $\mathbb{Q} - \{0\} < \mathbb{R} - \{0\}$ and $\mathbb{R} - \{0\} < \mathbb{C} - \{0\}$.

(iv) $\{+1, -1\} < C$ and $C < \mathbb{C} - \{0\}$.

(v) The elements e, r, r^2, r^3, r^4, r^5 form a subgroup of D_6. These are the rotations which leave the plate the same way up.

(vi) The collection e, r, s, rs does not form a subgroup of D_6. The element r lies in the set, but the product $rr = r^2$ is missing.

(vii) The integers 0, 2, 4 form a subgroup of $\mathbb{Z}_6$.

Example (v) generalises as follows: Given a group G and an element x of G, the set of all powers of x (i.e., the set of all elements of G of the form x^n for some integer n) is a subgroup of G. (The product $x^m x^n$ of two powers of x is x^{m+n}, which is again a power of x, the identity element of G is x^0, and the inverse of x^n is x^{-n}, which is also a power of x.) This subgroup is called the *subgroup generated by x* and written $\langle x \rangle$. If x has infinite order, then $\langle x \rangle$ consists of $\ldots x^{-2}$, x^{-1}, e, x, x^2, x^3, $\ldots$. If x has finite order, say m, the elements of $\langle x \rangle$ are e, x, x^2, $\ldots$, x^{m-1}. So the order of x is precisely the order of the subgroup generated by x. If there is an element x in G which generates all of G (in other words, for which $\langle x \rangle = G$), we say that G is a *cyclic group*.

EXAMPLES. (i) The number 1 generates $\mathbb{Z}$, as does -1, so $\mathbb{Z}$ is an infinite cyclic group. (Since the group structure is addition of integers, the fourth power, say, of 1 is $1 + 1 + 1 + 1 = 4$.)

(ii) The number 1 generates $\mathbb{Z}_n$, so $\mathbb{Z}_n$ is a cyclic group of order n.

(iii) In $\mathbb{Z}_6$ we have

$$\langle 0 \rangle = \{0\},$$
$$\langle 1 \rangle = \langle 5 \rangle = \mathbb{Z}_6,$$
$$\langle 2 \rangle = \langle 4 \rangle = \{0, 2, 4\},$$
$$\langle 3 \rangle = \{0, 3\}.$$

For example, the elements of $\langle 4 \rangle$ are 4, $4 +_6 4 = 2$, and $4 +_6 4 +_6 4 = 0$.

(iv) In D_3 we have

$$\langle e \rangle = \{e\},$$
$$\langle r \rangle = \langle r^2 \rangle = \{e, r, r^2\},$$
$$\langle s \rangle = \{e, s\},$$
$$\langle rs \rangle = \{e, rs\},$$
$$\langle r^2 s \rangle = \{e, r^2 s\}.$$

The dihedral group D_n is not cyclic, but each of its elements can be written in terms of the two elements r and s, and we say that r and s together generate D_n. Let X be a non-empty subset of a group G. An expression of the form

$$x_1^{m_1} x_2^{m_2} \ldots x_k^{m_k} \qquad (*)$$

where $x_1, \ldots, x_k$ belong to X (they need not be distinct) and $m_1, \ldots, m_k$ are integers is called a *word* in the elements of X. The collection of all these words is a *subgroup* of G. (Given two of them, writing one after the other shows that a product of two words in the elements of X is again a word in the elements of X. The identity element of G can be thought of as the word x^0 for any element x of X, and the inverse of the word

$$x_1^{m_1} x_2^{m_2} \ldots x_k^{m_k} \quad \text{is} \quad x_k^{-m_k} \ldots x_2^{-m_2} x_1^{-m_1}$$

which is also a word in the elements of X.) This subgroup is called the *subgroup generated by X*. If it fills out all of G we say that *X is a set of generators for G*, or that *the elements of X together generate G*. Suppose X is a set of generators for G, and let Y be another subset of G. Then if Y contains X it is also a set of generators for G. More generally, if every element of X can be written as a word in the elements of Y, then Y is a set of generators for G.

EXAMPLES. (i) The elements r and s together generate D_n. This choice of two generators is not unique. For example rs and s together also generate D_n, because $r = (rs)s$ and therefore any word in r and s can be converted into a word in rs and s.

(ii) The group structure on $\mathbb{C}$ is addition of complex numbers; consequently, words are written as linear combinations $m_1 z_1 + m_2 z_2 + \cdots + m_k z_k$ with

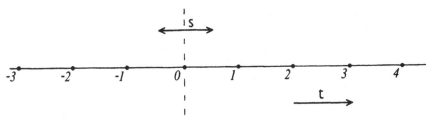

Figure 5.2

integer coefficients. The subgroup generated by $\{1, i\}$ is the group of Gaussian integers whose elements are the complex numbers $a + ib$ for which $a, b \in \mathbb{Z}$.

(iii) Consider the real line with the set of integers marked on as in Figure 5.2. Let G be the set of functions from the line to itself which preserve distance and which send the integers among themselves. Then G is a group under composition of functions. It is not hard to check (Exercise 5.9) that each element of G is either a translation to the left or right through an integral distance, a reflection in an integer point, or a reflection in a point which lies midway between two integers. Let t be translation to the right through one unit, so $t(x) = x + 1$, and let s be reflection in the origin, so $s(x) = -x$. Then the elements of G are

$$\ldots t^{-2}, t^{-1}, e, t, t^2, \ldots$$

$$\ldots t^{-2}s, t^{-1}s, s, ts, t^2s, \ldots \qquad (**)$$

where e is the identity function. For example $t^{-2}(x) = x - 2$, showing that t^{-2} is translation to the left through two units, and $ts(x) = t(-x) = -x + 1$, showing that ts is reflection in the point $\frac{1}{2}$. The translation t and the reflection s together generate G. Equally well the two reflections ts and s together generate G. Note that

$$st(x) = s(x + 1) = -x - 1$$

and

$$t^{-1}s(x) = t^{-1}(-x) = -x - 1,$$

which means $st = t^{-1}s$. Knowing $s^2 = e$ and $st = t^{-1}s$ allows us to multiply any two elements from the list $(**)$ and manipulate the product to have the same form. This reminds us very much of D_n. Indeed, the only difference is that the rotation r of order n has been replaced by the translation t of infinite order. For this reason we call G the **infinite dihedral group** and denote it by D_∞.

We end this section with one or two useful facts about subgroups.

(5.1) Theorem. *A non-empty subset H of a group G is a subgroup of G if and only if xy^{-1} belongs to H whenever x and y belong to H.*

Proof. If H is a subgroup, and if $x, y \in H$, then we know y^{-1} must be in H, and so the product xy^{-1} belongs to H. Conversely, suppose H is non-empty and that $xy^{-1} \in H$ whenever $x, y \in H$. If $x \in H$, then $e = xx^{-1} \in H$, and $x^{-1} = ex^{-1} \in H$. Finally, if y is also an element of H, then $y^{-1} \in H$ as above, and so $xy = x(y^{-1})^{-1} \in H$. Therefore, H is a subgroup of G.   □

(5.2) Theorem. *The intersection of two subgroups of a group is itself a subgroup.*

Proof. Let H and K be subgroups of the group G. The identity lies in both H and K, so $H \cap K$ is non-empty. If x and y are elements of the intersection $H \cap K$, they are both elements of H and both elements of K. Since H, K are subgroups the product xy^{-1} lies in H and K. Therefore $xy^{-1} \in H \cap K$ and we can apply (5.1).   □

(5.3) Theorem. (a) *Every subgroup of $\mathbb{Z}$ is cyclic.*
(b) *Even better, every subgroup of a cyclic group is cyclic.*

Proof of (a). Let H be a subgroup of $\mathbb{Z}$. If $H = \{0\}$ then H is cyclic. If $H \neq \{0\}$, then H contains a non-zero integer x, and since H is a subgroup it must also contain $-x$. So H contains a positive integer. Let d be the smallest positive integer in H. We claim that d generates H. If $n \in H$, divide n by d to give $n = qd + m$ where q and m are integers and $0 \leqslant m < d$. In other words $m = n \pmod{d}$. We know that $n \in H$ and $d \in H$. As H is a subgroup $qd \in H$, hence $-qd \in H$, and therefore

$$m = n - qd = n + (-qd)$$

belongs to H. This contradicts our choice of d unless m is zero. Consequently $n = qd$, showing each element of H to be an integer multiple of d as required.   □

Proof of (b). Let G be a cyclic group and K a subgroup of G which is not the trivial subgroup $\{e\}$. If x is a generator for G, then every element of G, and hence every element of K, is a power of x. Let $H = \{n \in \mathbb{Z} \mid x^n \in K\}$. One easily checks that H is a subgroup of $\mathbb{Z}$. By (a) H is cyclic and, if d generates H, then x^d generates K. This completes the proof.   □

EXERCISES

5.1. Find all the subgroups of each of the groups $\mathbb{Z}_4$, $\mathbb{Z}_7$, $\mathbb{Z}_{12}$, D_4, and D_5.

5.2. If m and n are positive integers, and if m is a factor of n, show that $\mathbb{Z}_n$ contains a subgroup of order m. Does $\mathbb{Z}_n$ contain more than one subgroup of order m?

5.3. With the notation of Section 4, check that rs and r^2s together generate D_n.

5.4. Find the subgroup of D_n generated by r^2 and r^2s, distinguishing carefully between the cases n odd and n even.

5.5. Suppose H is a *finite* non-empty subset of a group G. Prove that H is a subgroup of G if and only if xy belongs to H whenever x and y belong to H.

5.6. Draw a diagonal in a regular hexagon. List those plane symmetries of the hexagon which leave the diagonal fixed, and those which send the diagonal to itself. Show that both collections of symmetries are subgroups of the group of all plane symmetries of the hexagon.

5.7. Let G be an *abelian* group and let H consist of those elements of G which have finite order. Prove that H is a subgroup of G.

5.8. Which elements of the infinite dihedral group have finite order? Do these elements form a subgroup of D_∞?

5.9. Let f be a function from the real line to itself which preserves the distance between every pair of points and which sends the integers among themselves.

   (a) Assuming f has no fixed points, show that f is a *translation* through an integral distance.
   (b) If f leaves exactly one point fixed, show that this point is either an integer or lies midway between two integers, and that f is *reflection* in this fixed point.
   (c) Finally, check that f must be the identity if it leaves more than one point fixed.

5.10. Make a list of those elements of $\mathbb{Z}_{12}$ which generate $\mathbb{Z}_{12}$. Answer the same question for $\mathbb{Z}_5$ and for $\mathbb{Z}_9$. Do your answers suggest a general result?

5.11. Show that $\mathbb{Q}$ is not cyclic. Even better, prove that $\mathbb{Q}$ cannot be generated by a finite number of elements.

5.12. If $a, b \in \mathbb{Z}$ are not both zero and if $H = \{\lambda a + \mu b \mid \lambda, \mu \in \mathbb{Z}\}$, show that H is a subgroup of $\mathbb{Z}$. Let d be the smallest positive integer in H. Check that d is the highest common factor of a and b. (Consequently, the highest common factor of two integers a, b can always be written as a linear combination $\lambda a + \mu b$ with integer coefficients.)

CHAPTER 6

Permutations

We continue to increase our stock of examples by introducing groups of permutations. Rearranging or permuting the members of a set is a familiar idea, for example interchanging 1 and 3 while leaving 2 fixed gives a permutation of the first three integers. By a *permutation* of an arbitrary set X we shall mean a bijection from X to itself. The collection of *all* permutations of X forms a group S_X under composition of functions. There is very little to check. If $\alpha: X \to X$ and $\beta: X \to X$ are permutations, the composite function $\alpha\beta: X \to X$ defined by $\alpha\beta(x) = \alpha(\beta(x))$ is also a permutation. Composition of functions is associative, and the special permutation ε which leaves every point of X fixed clearly acts as an identity. Finally, each permutation α is a bijection and therefore has an inverse $\alpha^{-1}: X \to X$, which is also a permutation and which satisfies $\alpha^{-1}\alpha = \varepsilon = \alpha\alpha^{-1}$. If X is an infinite set, S_X is an infinite group. When X consists of the first n positive integers, then S_X is written S_n and called the *symmetric group* of degree n. The order of S_n is $n!$.

For the time being, we shall concentrate on the symmetric groups. Here are the six elements of S_3.

$$\varepsilon = \begin{bmatrix} 123 \\ 123 \end{bmatrix}, \begin{bmatrix} 123 \\ 213 \end{bmatrix}, \begin{bmatrix} 123 \\ 321 \end{bmatrix}, \begin{bmatrix} 123 \\ 132 \end{bmatrix}, \begin{bmatrix} 123 \\ 231 \end{bmatrix}, \begin{bmatrix} 123 \\ 312 \end{bmatrix}.$$

To find the image of an integer under a particular permutation just look vertically underneath it. Thus

$$\begin{bmatrix} 123 \\ 312 \end{bmatrix}$$

sends 1 to 3, 2 to 1, and 3 to 2. Remembering that $\alpha\beta$ *means first apply β, then apply α*, we calculate

$$\begin{bmatrix} 123 \\ 213 \end{bmatrix}\begin{bmatrix} 123 \\ 132 \end{bmatrix} = \begin{bmatrix} 123 \\ 231 \end{bmatrix},$$

whereas

$$\begin{bmatrix} 123 \\ 132 \end{bmatrix}\begin{bmatrix} 123 \\ 213 \end{bmatrix} = \begin{bmatrix} 123 \\ 312 \end{bmatrix}. \tag{$*$}$$

Therefore, S_3 is not abelian. We can immediately say that S_n is not abelian when $n \geqslant 3$. Why?

When extended to higher values of n, this notation is too cumbersome to work with. For example, the element α of S_6 defined by $\alpha(1) = 5$, $\alpha(2) = 4$, $\alpha(3) = 3$, $\alpha(4) = 6$, $\alpha(5) = 1$, $\alpha(6) = 2$ becomes

$$\alpha = \begin{bmatrix} 123456 \\ 543612 \end{bmatrix}.$$

The same information can be captured by $\alpha = (15)(246)$. Inside each pair of brackets an integer is sent to the integer following it, the final integer being sent to the first. Therefore 1 is sent to 5 and 5 to 1, 2 is sent to 4, 4 to 6, and 6 to 2. There is no need to mention integers which are left fixed by the permutation. Here there is no mention of 3. *We can describe any permutation in this way*, the prescription being as follows: Open a pair of brackets, then write down the smallest integer which is moved by the given permutation. Now, list the image of this integer under the permutation, followed by its image and so on, eventually closing the brackets at the stage where we would come full circle to our starting point. Open a new pair of brackets, list the smallest integer which has so far not been mentioned and which is moved by the permutation, etc.

EXAMPLES

(i) $\qquad \begin{bmatrix} 123456789 \\ 189362754 \end{bmatrix} = (2856)(394).$

(ii) $\qquad \begin{bmatrix} 12345678 \\ 81673542 \end{bmatrix} = (182)(365)(47).$

(iii) The elements of S_3 are

$$\varepsilon, \ (12), \ (13), \ (23), \ (123), \text{ and } (132).$$

(iv) The calculation ($*$) becomes

$$(12)(23) = (123), \qquad \text{whereas} \qquad (23)(12) = (132).$$

With this new notation a permutation $(a_1 a_2 \ldots a_k)$ inside a single pair of brackets is called a *cyclic permutation*. It sends a_1 to a_2, a_2 to $a_3, \ldots, a_{k-1}$ to a_k and a_k to a_1, leaving all other integers fixed. The number k is its length and a cyclic permutation of length k is called a *k-cycle*. A 2-cycle is usually referred to as a **transposition**. The above argument shows that *every element of S_n may be written as a product of disjoint cyclic permutations*, disjoint in the sense that no integer is moved by more than one of them.

Look again at Example (i) where we obtain (2856) and (394). The first of these affects only the integers 2, 5, 6, 8, and the second moves only 3, 4 and 9. Because these permutations are disjoint, they *commute* with one another, that is to say $(2856)(394) = (394)(2856)$. There is of course a general result here, if α and β are elements of S_n and if no integer is moved by both α and β then $\alpha\beta = \beta\alpha$. The decomposition of an element of S_n as a product of disjoint cyclic permutations is unique up to the order in which we write down these cyclic permutations.

(6.1) Theorem. *The transpositions in S_n together generate S_n.*

Proof. Each element of S_n can be written as a product of cyclic permutations, and any cyclic permutation can be written as a product of transpositions, since

$$(a_1 a_2 \ldots a_k) = (a_1 a_k) \ldots (a_1 a_3)(a_1 a_2).$$

Therefore, each element of S_n can be written as a product of transpositions. Note that these transpositions need not be disjoint, and that this decomposition is not unique. □

EXAMPLE

$$\begin{bmatrix} 123456 \\ 543612 \end{bmatrix} = (15)(246) = (15)(26)(24).$$

Since $(246) = (624)$ we have, equally well,

$$\begin{bmatrix} 123456 \\ 543612 \end{bmatrix} = (15)(624) = (15)(64)(62)$$
$$= (15)(46)(26).$$

(6.2) Theorem. (a) *The transpositions* $(12), (13), \ldots, (1n)$ *together generate S_n.*
(b) *The transpositions* $(12), (23), \ldots, (n-1n)$ *together generate S_n.*

Proof. (a) Note that $(ab) = (1a)(1b)(1a)$ and use (6.1).
(b) Note that $(1k) = (k-1k)\ldots(34)(23)(12)(23)(34)\ldots(k-1k)$ and use part (a). □

(6.3) Theorem. *The transposition* (12) *and the n-cycle* $(12 \ldots n)$ *together generate S_n.*

Proof. By (6.2)(b) we need only write each transposition of the form $(k\,k+1)$ as a word in (12) and $(12\ldots n)$. This can be achieved as follows

$$(23) = (12\ldots n)(12)(12\ldots n)^{-1},$$

and more generally

$$(kk + 1) = (12 \ldots n)^{k-1}(12)(12 \ldots n)^{1-k}$$

for $2 \leqslant k < n$.   □

A given element of S_n can be written as a product of transpositions in many different ways. *However, the number of transpositions which occur is either always even or always odd.* To check this we introduce the polynomial

$$P = P(x_1, x_2, \ldots, x_n)$$
$$= (x_1 - x_2)(x_1 - x_3) \ldots (x_1 - x_n)(x_2 - x_3) \ldots (x_{n-1} - x_n),$$

in other words, the product of all factors $(x_i - x_j)$ where $1 \leqslant i \leqslant n, 1 \leqslant j \leqslant n$, and $i < j$. If $\alpha \in S_n$, we define αP to be the product of all factors $(x_{\alpha(i)} - x_{\alpha(j)})$ where again $1 \leqslant i \leqslant n, 1 \leqslant j \leqslant n$, and $i < j$. The effect of α is to permute the terms of P, while at the same time changing the sign of some of them. Therefore, αP is either $+P$ or $-P$, and this determines the so called *sign* of α to be $+1$ in the first instance and -1 in the second. To clear the air, here is an example.

EXAMPLE. $n = 3$ and $\alpha = (132)$. Then

$$P = (x_1 - x_2)(x_1 - x_3)(x_2 - x_3),$$

and

$$\alpha P = (x_3 - x_1)(x_3 - x_2)(x_1 - x_2) = +P.$$

So the sign of (132) is $+1$.

In general, if $\alpha, \beta \in S_n$ the sign of $\alpha\beta$ is the product of the signs of α and β, and the sign of the transposition (12) is clearly -1. Since $(1a) = (2a)(12)(2a)$ for $a > 2$, the sign of $(1a)$ is also -1, and since $(ab) = (1a)(1b)(1a)$, the sign of any transposition is -1. Consequently, if an element of S_n can be written as the product of an even number of transpositions, then its sign must be $+1$, whereas the product of an odd number of transpositions always has sign -1.

An element of S_n which can be expressed as the product of an even number of transpositions is called an **even permutation**: the others are **odd permutations**. Since

$$(a_1 a_2 \ldots a_k) = (a_1 a_k) \ldots (a_1 a_3)(a_1 a_2),$$

a cyclic permutation is even precisely when its length is odd.

(6.4) Theorem. *The even permutations in S_n form a subgroup of order $n!/2$ called the **alternating group** A_n of degree n.*

Proof. If α and β are even permutations, write each of them as the product of an even number of transpositions. Juxtaposition of these products shows that

$\alpha\beta$ is even. Writing the product for α in the reverse order shows that α^{-1} is even. The identity is of course even because $\varepsilon = (12)(12)$. Therefore, the even permutations form a subgroup of S_n. If α is even then $(12)\alpha$ is odd. This pairs off the elements of S_n and shows that precisely half its elements are even. (Why can *every* odd permutation be expressed as an even permutation followed by (12)?)   □

(6.5) Theorem. *For* $n \geqslant 3$ *the 3-cycles generate* A_n.

Proof. Each 3-cycle is certainly an even permutation. Given an element of A_n, use (6.2a) to write it as the product of an even number of transpositions of the form $(1a)$. Collect these transpositions into adjacent *pairs*, and combine each pair using $(1a)(1b) = (1ba)$. Our element is now expressed as a product of 3-cycles.   □

The twelve elements of A_4 are

$$\varepsilon, \qquad (12)(34), \qquad (13)(24), \qquad (14)(23)$$
$$(123), \qquad (124), \qquad (134), \qquad (234),$$
$$(132), \qquad (142), \qquad (143), \qquad (243).$$

The remaining elements of S_4, the odd permutations, are

$$(12), (13), (14), (23), (24), (34),$$
$$(1234), \qquad (1243), \qquad (1324),$$
$$(1432), \qquad (1342), \qquad (1423).$$

If we wish to write $(13)(24)$ as a product of 3-cycles, the procedure in (6.5) gives

$$(13)(24) = (13)(12)(14)(12)$$
$$= (123)(124).$$

EXERCISES

6.1. Write out a multiplication table for S_3.

6.2. Express each of the following elements of S_8 as a product of disjoint cyclic permutations, and as a product of transpositions.

(a) $\begin{bmatrix} 12345678 \\ 76418235 \end{bmatrix}$ (b) $(4568)(1245)$

(c) $(624)(253)(876)(45)$

Which, if any, of these permutations belong to A_8?

6.3. Show that the elements of S_9 which send the numbers 2, 5, 7 among themselves form a subgroup of S_9. What is the order of this subgroup?

6.4. Find a subgroup of S_4 which contains six elements. How many subgroups of order six are there in S_4?

6.5. Compute $\alpha P(x_1, x_2, x_3, x_4)$ when $\alpha = (143)$ and when $\alpha = (23)(412)$.

6.6. If H is a subgroup of S_n, and if H is not contained in A_n, prove that precisely one-half of the elements of H are even permutations.

6.7. Show that if n is at least 4 every element of S_n can be written as a product of two permutations, each of which has order 2. (Experiment first with cyclic permutations.)

6.8. If α, β are elements of S_n, check that $\alpha\beta\alpha^{-1}\beta^{-1}$ always lies in A_n, and that $\alpha\beta\alpha^{-1}$ belongs to A_n whenever β is an even permutation. Work out these elements when $n = 4$, $\alpha = (2143)$, and $\beta = (423)$.

6.9. When n is odd show that (123) and $(12 \ldots n)$ together generate A_n. If n is even show that (123) and $(23 \ldots n)$ together generate A_n.

6.10. If $\alpha, \beta \in S_n$ and if $\alpha\beta = \beta\alpha$, prove that β permutes those integers which are left fixed by α. Show that β must be a power of α when α is an n-cycle.

6.11. Find the order of each permutation listed in Exercise 6.2.

6.12. Prove that the order of an element α of S_n is the least common multiple of the lengths of the cycles which are obtained when α is written as a product of disjoint cyclic permutations.

CHAPTER 7

Isomorphisms

A chessboard (Fig. 7.1) has four plane symmetries, the identity e, rotation r through π about its centre, and the reflections q_1, q_2 in its two diagonals. They form a group under composition whose multiplication table is given below. It is easy to check that multiplication modulo eight makes the numbers 1, 3, 5, 7 into a group. Again we provide the corresponding table.

	e	r	q_1	q_2
e	e	r	q_1	q_2
r	r	e	q_2	q_1
q_1	q_1	q_2	e	r
q_2	q_2	q_1	r	e

	1	3	5	7
1	1	3	5	7
3	3	1	7	5
5	5	7	1	3
7	7	5	3	1

There is an apparent similarity between these tables if we ignore their origins. In each case the group has four elements, and these elements appear to combine in the same manner. Only the way in which the elements are labelled distinguishes one table from the other.

Label the first group G, the second G', and the correspondence

$$e \to 1, \quad r \to 3, \quad q_1 \to 5, \quad q_2 \to 7$$

by $x \to x'$. When we say that the elements combine in the same manner, we mean that if $x \to x'$ and $y \to y'$, then $xy \to x'y'$. This correspondence is called an isomorphism between G and G'. It is a bijection and it carries the multiplication of G to that of G'. To all intents and purposes then, G and G' are "the same". Technically they are isomorphic in the following sense.

*Two groups G and G' are **isomorphic** if there is a bijection φ from G to G' which satisfies $\varphi(xy) = \varphi(x)\varphi(y)$ for all $x, y \in G$. The function φ is called an **isomorphism** between G and G'.*

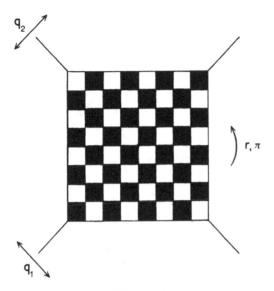

q_2

r, π

q_1

Figure 7.1

Asking for a bijection φ *from* G to G' ensures that the underlying sets of G and G' have the same size. If, in addition, $\varphi(xy) = \varphi(x)\varphi(y)$ for any two elements x, y of G, it does not matter whether we first combine two elements in G and then send their product into G' using φ, or first send the elements separately into G' via φ and then combine their images in G'. Therefore, G' is really just G in disguise. Notice that the inverse function $\varphi^{-1} \colon G' \to G$ is equally well an isomorphism, so our definition is symmetrical in G and G'. To indicate that two groups G and G' are isomorphic, we shall write $G \cong G'$.

EXAMPLES. (i) Define $\varphi \colon \mathbb{R} \to \mathbb{R}^{\text{pos}}$ by $\varphi(x) = e^x$. Then φ is a bijection and

$$\varphi(x + y) = e^{x+y} = e^x e^y = \varphi(x)\varphi(y)$$

for all $x, y \in \mathbb{R}$. Therefore $\mathbb{R}$ and $\mathbb{R}^{\text{pos}}$ are isomorphic groups. Remember that the group operation in $\mathbb{R}$ is addition, whereas that in $\mathbb{R}^{\text{pos}}$ is multiplication.

(ii) We already know a good deal about the tetrahedron. It has twelve rotational symmetries which form a non-abelian group G. We can learn more as follows. Number the vertices 1, 2, 3, 4 as in Figure 7.2. Each rotational symmetry induces a permutation of the vertices, and therefore a permutation of the first four integers. For example, the rotation r illustrated induces the cyclic permutation (234), and s induces (14)(23). Working systematically through all the other possibilities produces the twelve elements of A_4. If two rotations u, v induce permutations α, β, respectively then uv clearly induces $\alpha\beta$. Therefore, the correspondence

rotational symmetry $\to$ induced permutation

shows that G is isomorphic to A_4.

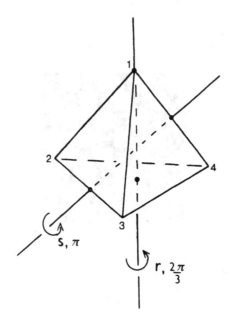

Figure 7.2

(iii) Any infinite cyclic group is isomorphic to $\mathbb{Z}$. If G is an infinite cyclic group, and if x generates G, define $\varphi: G \to \mathbb{Z}$ by $\varphi(x^m) = m$. Then φ is a bijection and

$$\varphi(x^m x^n) = \varphi(x^{m+n}) = m + n = \varphi(x^m) + \varphi(x^n).$$

This shows that φ is an isomorphism.

(iv) Any finite cyclic group of order n is isomorphic to $\mathbb{Z}_n$. If G is a cyclic group of order n, and if x generates G, define $\varphi: G \to \mathbb{Z}_n$ by $\varphi(x^m) = m \pmod{n}$. Then φ is an isomorphism.

(v) The numbers $1, -1, i, -i$ form a group under complex multiplication. It is cyclic and $i, -i$ are both generators. The procedure in (iv) gives two isomorphisms

$$1 \to 0, \qquad i \to 1, \quad -1 \to 2, \quad -i \to 3, \quad \text{and}$$
$$1 \to 0, \quad -i \to 1, \quad -1 \to 2, \qquad i \to 3$$

between this group and $\mathbb{Z}_4$.

(vi) D_3 and S_3 are isomorphic. Do this in the spirit of the calculation for the tetrahedron by labelling the vertices of an equilateral triangle 1, 2, 3.

(vii) There is no isomorphism between $\mathbb{Q}$ and $\mathbb{Q}^{\text{pos}}$. For suppose $\varphi: \mathbb{Q} \to \mathbb{Q}^{\text{pos}}$ is a candidate. Choose $x \in \mathbb{Q}$ such that $\varphi(x) = 2$, then

$$\varphi\left[\frac{x}{2}+\frac{x}{2}\right]=\varphi\left[\frac{x}{2}\right]\varphi\left[\frac{x}{2}\right]=2$$

and $\varphi(x/2)$ has to be $\sqrt{2}$. Since $\sqrt{2}$ is irrational, we have a contradiction.

An isomorphism $\varphi: G \to G'$ is a bijection, therefore G and G' must have the same order. It sends the identity of G to that of G'. For suppose $x' \in G'$ and $\varphi(x) = x'$, then

$$x'\varphi(e) = \varphi(x)\varphi(e) = \varphi(xe) = \varphi(x) = x'$$

and similarly $\varphi(e)x' = x'$, showing that $\varphi(e)$ is the identity element of G'. Alternatively we can observe that

$$\varphi(e)\varphi(e) = \varphi(ee) = \varphi(e)$$

and then multiply both sides of this equation by the inverse of $\varphi(e)$ in G'. The latter argument is preferable to the first one; it does not use the fact that φ is a bijection, only that φ sends the multiplication of G to the multiplication of G'.

The isomorphism φ sends inverses to inverses in the sense that

$$\varphi(x)^{-1} = \varphi(x^{-1}) \qquad \text{for all } x \in G.$$

Again this is easy to check. We have

$$\varphi(x^{-1})\varphi(x) = \varphi(x^{-1}x) = \varphi(e) = e,$$

and similarly $\varphi(x)\varphi(x^{-1}) = e$. Therefore $\varphi(x^{-1})$ is indeed the inverse of $\varphi(x)$ in G'. *If G is abelian then so is G'.* For if $x',y' \in G'$ and if $\varphi(x) = x'$, $\varphi(y) = y'$ then

$$x'y' = \varphi(x)\varphi(y) = \varphi(xy)$$

$$= \varphi(yx) \qquad \text{because } G \text{ is abelian}$$

$$= \varphi(y)\varphi(x) = y'x'.$$

If $\varphi: G \to G'$ is an isomorphism and if H is a subgroup of G, then $\varphi(H)$ is a subgroup of G'. We check this using (5.1). Suppose x',y' are elements of $\varphi(H)$, then we can find $x,y \in H$ such that $\varphi(x) = x'$ and $\varphi(y) = y'$. Now H is a subgroup of G, so xy^{-1} belongs to H, and since

$$\varphi(xy^{-1}) = \varphi(x)\varphi(y^{-1}) = \varphi(x)\varphi(y)^{-1} = x'(y')^{-1}$$

we see that $x'(y')^{-1}$ belongs to $\varphi(H)$ as required. Consider the special case where H is cyclic, generated by the element g of G. If $x' \in \varphi(H)$, we have $x' = \varphi(g^m) = \varphi(g)^m$ for some integer m. Therefore, $\varphi(H)$ is generated by the element $\varphi(g)$. As H and $\varphi(H)$ have the same number of elements, the order of $\varphi(g)$ must be the same as the order of g. *Therefore, an isomorphism preserves the order of each element.* We comment, finally, that if $\varphi: G \to G'$ and $\psi: G' \to G''$ are both isomorphisms, then the composition $\psi\varphi: G \to G''$ is also an isomorphism.

In Chapter 1 we introduced three solids; a regular tetrahedron, a flat hexagonal plate with equal sides, and a right regular prism on a twelve-sided base. The *geometrical observation* that these solids exhibit different amounts of symmetry translates to the *algebraic statement* that no two of them have isomorphic symmetry groups. We have just shown the rotational symmetry group of the tetrahedron to be isomorphic to A_4. That of the plate is by definition D_6, and that of the pyramid is cyclic (generated by a rotation through $\pi/6$ about the single axis of symmetry) and must therefore be isomorphic to $\mathbb{Z}_{12}$. Of these three groups $\mathbb{Z}_{12}$ is the only one which is abelian, so it cannot be isomorphic to either of the other two. And D_6, unlike A_4, contains an element of order six, so D_6 cannot be isomorphic to A_4.

EXERCISES

7.1. Check that the numbers 1, 2, 4, 5, 7, 8 form a group under multiplication modulo 9 and show that this group is isomorphic to $\mathbb{Z}_6$.

7.2. Verify that the integers 1, 3, 7, 9, 11, 13, 17, 19 form a group under multiplication modulo 20. Explain why this group is not isomorphic to $\mathbb{Z}_8$.

7.3. Show that the subgroup $\{\varepsilon, (12)(34), (13)(24), (14)(23)\}$ of A_4 is isomorphic to the group of plane symmetries of a chessboard.

7.4. Produce a specific isomorphism between S_3 and D_3. How many different isomorphisms are there from S_3 to D_3?

7.5. Let G be a group. Show that the correspondence $x \leftrightarrow x^{-1}$ is an isomorphism from G to G if and only if G is *abelian*.

7.6. Prove that $\mathbb{Q}^{\text{pos}}$ is not isomorphic to $\mathbb{Z}$.

7.7. If G is a group, and if g is an element of G, show that the function $\varphi: G \to G$ defined by $\varphi(x) = gxg^{-1}$ is an isomorphism. Work out this isomorphism when G is A_4 and g is the permutation (123).

7.8. Call H a **proper subgroup** of the group G if H is neither $\{e\}$ nor all of G. Find a group which is isomorphic to one of its proper subgroups.

7.9. Suppose G is a cyclic group. If x generates G, and if $\varphi: G \to G$ is an isomorphism, prove that φ is completely determined by $\varphi(x)$ and that $\varphi(x)$ also generates G. Use these facts to find all isomorphisms from $\mathbb{Z}$ to $\mathbb{Z}$, and all isomorphisms from $\mathbb{Z}_{12}$ to $\mathbb{Z}_{12}$.

7.10. Show that $\mathbb{R}$ is not isomorphic to $\mathbb{Q}$ and that $\mathbb{R} - \{0\}$ is not isomorphic to $\mathbb{Q} - \{0\}$. Is $\mathbb{R}$ isomorphic to $\mathbb{R} - \{0\}$?

7.11. Prove that the subgroup of S_6 generated by (1234) and (56) is isomorphic to the group described in Exercise 7.2.

7.12. Show that the subgroup of S_4 generated by (1234) and (24) is isomorphic to D_4.

CHAPTER 8

Plato's Solids and Cayley's Theorem

There are five convex regular solids, the *tetrahedron* (four triangular faces), *cube* (six square faces), *octahedron* (eight triangular faces), *dodecahedron* (twelve pentagonal faces), and *icosahedron* (twenty triangular faces). They are illustrated in Figure 8.1. We have already shown that the group of rotational symmetries of the tetrahedron is isomorphic to the alternating group A_4. In this chapter we shall produce analogous results for the other four solids.

A *cube* has twenty-four rotational symmetries. They may be counted in the same way as for the tetrahedron, by finding all axes of symmetry together with the number of distinct rotations about each axis. The different types of axis are represented by L, M, and N in Figure 8.2. There are three axes such as L which together provide a total of nine rotations, six axes of type M with just one rotation each, and four principal diagonals like N about each of which the cube can be rotated through $2\pi/3$ and $4\pi/3$. This accounts for all twenty-three non-identity symmetries.

By numbering the vertices of the cube, we could produce an isomorphism from the cube group to a subgroup of S_8. A much better observation is that each rotational symmetry permutes the four principal diagonals of the cube. We shall use this fact to show that our group is isomorphic to S_4. Label the corners of the cube as shown, and let N_k denote the diagonal which joins corner k to corner k', $1 \leqslant k \leqslant 4$. Each rotational symmetry permutes N_1, N_2, N_3, N_4 among themselves and consequently gives us a permutation of the numbers 1, 2, 3, 4. For example, referring to Figure 8.2, r sends N_1 to N_2, N_2 to N_3, N_3 to N_4 and N_4 to N_1, giving the 4-cycle (1234), and s induces (143). Under t, diagonals N_1 and N_2 are interchanged, while N_3 and N_4 are sent to themselves (though they are not left fixed, each has its ends swapped round), so t gives the transposition (12). Write G for the symmetry group and $\varphi\colon G \to S_4$ for the function constructed above. Since the product of two rotations

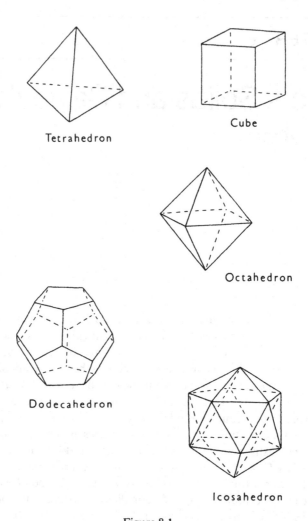

Tetrahedron

Cube

Octahedron

Dodecahedron

Icosahedron

Figure 8.1

clearly induces the appropriate product permutation, it only remains to check that φ is a bijection.

This can of course be done by elimination; in other words, by working out the effect of every single rotation of the cube on the four diagonals, rather a tedious process. Instead, we remember that a surjection between two finite sets which have the same number of elements must be a bijection, and show that φ is surjective. But this is easy. Both (1234) and (12) lie in $\varphi(G)$, and $\varphi(G)$ is a subgroup of S_4 because φ sends the multiplication of G to that of S_4. Therefore, every word which can be formed from (1234) and (12) must belong to $\varphi(G)$. Since (1234) and (12) together generate *all* of S_4, we have $\varphi(G) = S_4$, completing our argument.

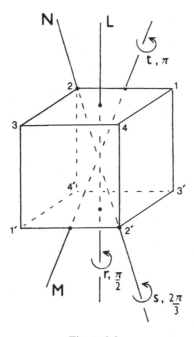

Figure 8.2

By joining up the centres of each pair of adjacent faces of a cube we can produce a regular *octahedron* inscribed in the cube (Fig. 8.3). The same procedure carried out on an octahedron gives a cube inscribed in the octahedron, and we say that the cube and the octahedron are **dual** solids. They clearly have the same amount of symmetry. Any symmetry of the cube is a symmetry of the inscribed dual octahedron, and vice versa. Without further ado we can say that the rotational symmetry groups of the cube and the octahedron are isomorphic.

There are two more regular solids, the *dodecahedron* and the *icosahedron*. They are dual to one another, and the reader should check this; so for the purposes of rotational symmetry we need only examine one of them, say the

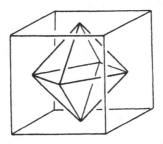

Figure 8.3

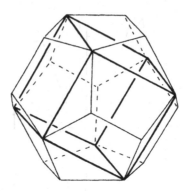

Figure 8.4

dodecahedron. Figure 8.4 shows a cube inside a dodecahedron. Each vertex of the cube is a vertex of the dodecahedron, and each edge is a diagonal of one of the pentagonal faces. If we look at a particular pentagon, exactly one of its five diagonals is an edge of the cube. There is nothing special about this diagonal and of course there are four more inscribed cubes corresponding to the other four diagonals of the pentagon. These five cubes are permuted by every rotational symmetry of the dodecahedron. Using them, just as we used the principal diagonals when looking at symmetries of a cube, it is not hard to check that the rotational symmetry group of a regular dodecahedron is isomorphic to A_5. We suggest the following steps and leave the details to the reader.

(i) Count the rotational symmetries of the dodecahedron and show there are sixty of them.
(ii) Observe that the order of A_5 is 60.
(iii) Number the inscribed cubes mentioned above 1 to 5 so that each rotation of the dodecahedron produces an element of S_5.
(iv) By considering rotations about axes which join opposite pairs of vertices, show that every 3-cycle in S_5 is produced in this way.
(v) Remember that the 3-cycles in S_5 together generate the subgroup A_5.

We summarise our results as follows. *The rotational symmetry group of the tetrahedron is isomorphic to A_4. The cube and octahedron both have rotational symmetry groups which are isomorphic to S_4. The dodecahedron and icosahedron both have rotational symmetry groups which are isomorphic to A_5.* We emphasise that all these groups contain only rotations. The full symmetry group of the tetrahedron is worked out in the exercises below; those of the other regular solids will be dealt with in Chapter 10.

So far we have represented the symmetry groups of the regular solids as groups of permutations. We now show that *every* group is isomorphic to a subgroup of a group of permutations.

(8.1) Cayley's Theorem. *Let G be a group, then G is isomorphic to a subgroup of S_G.*

Proof. Each element g in G gives a permutation $L_g: G \to G$ defined by $L_g(x) = gx$. (L_g is injective because if $L_g(x) = L_g(y)$ then $gx = gy$, giving $g^{-1}gx = g^{-1}gy$ and $x = y$. It is also surjective since if $z \in G$ then $L_g(g^{-1}z) = gg^{-1}z = z$.) We call L_g *left translation* by g. Notice that if $G = \mathbb{R}$ then L_g really is physical translation through a distance g. Let G' denote the subset $\{L_g | g \in G\}$ of S_G. Multiplication in S_G is composition of functions and

$$L_g(L_h(x)) = L_g(hx) = ghx = L_{gh}(x)$$

for all $x \in G$. Therefore, the product of two elements of G' lies in G'. The identity element ε of S_G belongs to G' because it equals L_e, and the inverse of L_g in S_G is $L_{g^{-1}}$ which also lies in G'. This shows that G' is a *subgroup* of S_G. The correspondence between G and G' defined by $g \to L_g$ is certainly surjective, and it sends the multiplication of G to that of G' because $gh \to L_{gh} = L_g L_h$. It is injective since if $L_g = L_h$, then $g = L_g(e) = L_h(e) = h$. Therefore, we have constructed an isomorphism between G and the subgroup G' of S_G.   □

(8.2) Theorem. *If G is a finite group of order n, then G is isomorphic to a subgroup of S_n.*

Proof. If the elements of G are numbered $1, 2, \ldots, n$ in some way, then each permutation of G induces a permutation of $1, 2, \ldots, n$. This gives an isomorphism from S_G to S_n and the subgroup G' of S_G is therefore isomorphic to a subgroup G'' of S_n. As G is isomorphic to G', and as the composition of two isomorphisms is an isomorphism, G is isomorphic to G''.   □

As an example we work out G'' when G is the group of plane symmetries of a chessboard introduced at the beginning of the previous chapter. From the multiplication table we have

$$L_r(e) = r, \qquad\qquad L_r(r) = r^2 = e,$$
$$L_r(q_1) = rq_1 = q_2, \qquad L_r(q_2) = rq_2 = q_1.$$

Therefore, L_r interchanges e and r, and interchanges q_1 and q_2. Calculating L_{q_1}, L_{q_2} in the same way, then labelling the elements e, r, q_1, q_2 with $1, 2, 3, 4$ respectively shows that G is isomorphic to the subgroup $\{\varepsilon, (12)(34), (13)(24), (14)(23)\}$ of S_4.

EXERCISES

8.1. Label the edges of a regular tetrahedron 1 to 6, so that each rotational symmetry of the tetrahedron produces an element of S_6. Work out the twelve elements of S_6 which occur in this way and check that they form a subgroup of S_6.

8.2. Join up the centres of each pair of adjacent faces of a regular tetrahedron and observe that the result is a second regular tetrahedron inscribed "upside down" in the first.

8.3. Number the faces of a cube 1 to 6. Find the elements of S_6 which correspond to the rotations r, s, and t of Figure 8.2.

8.4. Refer to the cube with a stripe marked on each face shown in Figure 18.1. Which rotational symmetries of the cube send stripes to stripes? To which subgroup of S_4 do these rotations correspond?

8.5. Draw five dodecahedra, with a different inscribed cube in each one. Use your pictures to examine how these cubes are permuted by the following rotations.

   (a) A rotation of the dodecahedron through $2\pi/5$ about an axis which joins the centres of a pair of opposite faces.
   (b) A rotation through π about an axis determined by the midpoints of a pair of opposite edges.
   (c) A rotation through $2\pi/3$ about an axis which joins a pair of opposite vertices.

8.6. Carry out the procedure of Cayley's theorem to obtain a subgroup of S_6 which is isomorphic to D_3.

8.7. Show that Cayley's theorem, when applied to $\mathbb{R}$, produces the subgroup of $S_{\mathbb{R}}$ which contains all translations of the real line.

8.8. Convert each element α of S_n into an element α_* of S_{n+2} as follows. The new permutation α_* behaves just like α on the integers $1, 2, \ldots, n$. If α is an even permutation then α_* fixes $n + 1$ and $n + 2$, whereas if α is odd α_* interchanges $n + 1$ and $n + 2$. Verify that α_* is always an *even* permutation and that the correspondence $\alpha \to \alpha_*$ defines an isomorphism from S_n to a subgroup of A_{n+2}. Work out this subgroup when $n = 3$.

8.9. If G is a finite group of order n, prove that G is isomorphic to a subgroup of the *alternating group* A_{n+2}.

8.10. If α is an element of S_n, write $\alpha_\#$ for the permutation in S_{2n} defined by

$$\alpha_\#(k) = \begin{cases} \alpha(k), & 1 \leqslant k \leqslant n \\ \alpha(k - n) + n, & n + 1 \leqslant k \leqslant 2n. \end{cases}$$

Show that $\alpha_\#$ is always an even permutation and that the correspondence $\alpha \to \alpha_\#$ is an isomorphism from S_n to a subgroup of A_{2n}. Work out this subgroup when $n = 3$.

8.11. Let G denote the *full* symmetry group of a regular tetrahedron T, and adopt the notation of Figure 7.2. Find a symmetry q of T which induces the transposition (12) of the vertices, and show that qr induces the 4-cycle (1234). Check that qr is neither a rotation nor a reflection, but *is* the product of three reflections. Count the symmetries of T and prove that G is isomorphic to S_4.

8.12. Working with the full symmetry group of the cube, show that each permutation of the principal diagonals can be realized by precisely two symmetries.

CHAPTER 9

Matrix Groups

The set of all invertible $n \times n$ matrices with real numbers as entries forms a group under matrix multiplication. We recall that if $A = (a_{ij})$, $B = (b_{ij})$ are two such matrices, the ijth entry of the *product* AB is the sum

$$a_{i1}b_{1j} + a_{i2}b_{2j} + \cdots + a_{in}b_{nj}.$$

Matrix multiplication is associative, the $n \times n$ identity matrix I_n plays the role of identity element, and the above product AB is invertible with inverse $B^{-1}A^{-1}$.

Each matrix A in this group determines an invertible linear transformation $f_A: \mathbb{R}^n \to \mathbb{R}^n$ defined by $f_A(\mathbf{x}) = \mathbf{x}A^t$ for all vectors $\mathbf{x} = (x_1, \ldots, x_n)$ in $\mathbb{R}^n$, where t stands for transpose. Since

$$f_{AB}(\mathbf{x}) = \mathbf{x}(AB)^t = \mathbf{x}B^t A^t = f_A(f_B(\mathbf{x}))$$

we see that *the product matrix AB determines the composite linear transformation $f_A f_B$.* Conversely, if $f: \mathbb{R}^n \to \mathbb{R}^n$ is an invertible linear transformation, and if A is the matrix which represents it with respect to the standard basis in both copies of $\mathbb{R}^n$, then A is invertible and $f = f_A$. For these reasons the group is called the **General Linear Group**, GL_n. If we wish to emphasise that the matrices all have real entries, then we write $GL_n(\mathbb{R})$. Changing $\mathbb{R}$ to $\mathbb{C}$ gives the corresponding group $GL_n(\mathbb{C})$ of $n \times n$ invertible complex matrices.

Matrix multiplication is not commutative for $n \geqslant 2$, so we have a family of *infinite non-abelian* groups $GL_2, GL_3 \ldots$. When $n = 1$ each matrix has a single entry which is a non-zero real number (non-zero because the matrix is invertible), and multiplication of matrices reduces to ordinary multiplication of numbers. Hence GL_1 is isomorphic to $\mathbb{R} - \{0\}$. If $A \in GL_n$, the $(n+1) \times (n+1)$ matrix

$$\tilde{A} = \begin{bmatrix} A & 0 \\ 0 & 1 \end{bmatrix}$$

belongs to GL_{n+1}. To construct $\tilde{A}$ we first add an extra column of zeros to A to produce an $n \times (n+1)$ matrix, then add to this an extra row, all of whose entries are zero except the final one which is 1. The collection of all matrices formed in this way is a *subgroup* of GL_{n+1} and the correspondence $A \to \tilde{A}$ shows that GL_n is isomorphic to this subgroup. In terms of linear transformations, if we identify $\mathbb{R}^n$ with the subspace of $\mathbb{R}^{n+1}$ consisting of those vectors which have zero final coordinate, then $f_{\tilde{A}}$ acts as f_A on $\mathbb{R}^n$ and leaves the last coordinate of each point unchanged. That is to say, $\mathbb{R}^{n+1} = \mathbb{R}^n \times \mathbb{R}$ and $f_{\tilde{A}} : \mathbb{R}^{n+1} \to \mathbb{R}^{n+1}$ is given by

$$f_{\tilde{A}}(\mathbf{x}, z) = (f_A(\mathbf{x}), z).$$

An $n \times n$ matrix A is **orthogonal** if $A^t A$ is the identity matrix, in which case the sum

$$a_{1i}a_{1j} + a_{2i}a_{2j} + \cdots + a_{ni}a_{nj}$$

is one when i equals j and zero otherwise. Therefore, the columns of A all have unit length and are perpendicular to one another; in technical language they form an orthonormal basis for $\mathbb{R}^n$. Of course, the rows of A also form an orthonormal basis for $\mathbb{R}^n$, and since $\det(A^t A) = (\det A)^2$, the determinant of A is either $+1$ or -1. If A and B are orthogonal, then

$$(AB^{-1})^t AB^{-1} = (B^{-1})^t A^t AB^{-1}$$
$$= (B^t)^t A^t AB^{-1}$$
$$= BA^t AB^{-1} = I_n.$$

Therefore, AB^{-1} is orthogonal and by Theorem (5.1) the collection of all $n \times n$ orthogonal matrices is a subgroup of GL_n. This subgroup is called the **Orthogonal Group**, O_n. Those elements of O_n which have determinant equal to $+1$ form a subgroup of O_n called the **Special Orthogonal Group**, SO_n.

If $A \in O_n$, the corresponding linear transformation f_A *preserves distance and preserves orthogonality*. To see why, let $\mathbf{x}$, $\mathbf{y}$ be points of $\mathbb{R}^n$ and consider the scalar product of $f_A(\mathbf{x})$ and $f_A(\mathbf{y})$. We have

$$f_A(\mathbf{x}) \cdot f_A(\mathbf{y}) = (\mathbf{x}A^t)(\mathbf{y}A^t)^t$$
$$= \mathbf{x}A^t A\mathbf{y}^t$$
$$= \mathbf{x}\mathbf{y}^t = \mathbf{x} \cdot \mathbf{y}$$

Since $\|\mathbf{x}\| = \sqrt{\mathbf{x} \cdot \mathbf{x}}$, setting $\mathbf{y}$ equal to $\mathbf{x}$ shows that $\|f_A(\mathbf{x})\| = \|\mathbf{x}\|$, so f_A preserves length. Also,

$$\|f_A(\mathbf{x}) - f_A(\mathbf{y})\| = \|f_A(\mathbf{x} - \mathbf{y})\| = \|\mathbf{x} - \mathbf{y}\|,$$

which shows that f_A preserves the distance between any two points. Finally, we note that $f_A(\mathbf{x}) \cdot f_A(\mathbf{y})$ is zero precisely when $\mathbf{x} \cdot \mathbf{y}$ is zero, therefore, if $\mathbf{x}$ and $\mathbf{y}$ are perpendicular vectors, then so are $f_A(\mathbf{x})$ and $f_A(\mathbf{y})$.

We can argue in the opposite direction. Suppose $f: \mathbb{R}^n \to \mathbb{R}^n$ is a linear transformation which preserves length. Then f preserves the distance between any two points because

$$\| f(\mathbf{x}) - f(\mathbf{y}) \| = \| f(\mathbf{x} - \mathbf{y}) \| = \| \mathbf{x} - \mathbf{y} \|,$$

and preserves right angles because

$$f(\mathbf{x}) \cdot f(\mathbf{y}) = \tfrac{1}{2}[\| f(\mathbf{x}) \|^2 - \| f(\mathbf{x}) - f(\mathbf{y}) \|^2 + \| f(\mathbf{y}) \|^2]$$

$$= \tfrac{1}{2}[\| \mathbf{x} \|^2 - \| \mathbf{x} - \mathbf{y} \|^2 + \| \mathbf{y} \|^2]$$

$$= \mathbf{x} \cdot \mathbf{y}.$$

Therefore, f maps the standard basis for $\mathbb{R}^n$ to an orthonormal basis. The matrix A which represents f has the elements of this basis as its columns; consequently A is orthogonal. We conclude that $f = f_A$ where $A \in O_n$.

We immediately feel more "at home" if we set n equal to 2 or 3. If $A \in O_2$ the columns of A are unit vectors and are orthogonal to one another. Suppose

$$A = \begin{bmatrix} a & c \\ b & d \end{bmatrix}$$

then (a, b) lies on the unit circle giving $a = \cos \theta$, $b = \sin \theta$ for some θ satisfying $0 \leqslant \theta < 2\pi$. As (c, d) is at right angles to (a, b) and also lies on the unit circle, we have $c = \cos \varphi$, $d = \sin \varphi$ where either $\varphi = \theta + \pi/2$ or $\varphi = \theta - \pi/2$. In the first case we obtain

$$\begin{bmatrix} \cos \theta & -\sin \theta \\ \sin \theta & \cos \theta \end{bmatrix},$$

which is an element of SO_2 and represents anticlockwise *rotation* through θ. The second case gives

$$\begin{bmatrix} \cos \theta & \sin \theta \\ \sin \theta & -\cos \theta \end{bmatrix}$$

which has determinant -1 and represents *reflection* in a line at angle $\theta/2$ to the positive x-axis. *Therefore, a* 2×2 *orthogonal matrix represents either a rotation of the plane about the origin, or a reflection in a straight line through the origin, and the matrix has determinant* $+1$ *precisely when it represents a rotation.* We have seen SO_2 before, albeit disguised as the unit circle in the complex plane. Each point on the unit circle has the form $e^{i\theta}$, where $0 \leqslant \theta < 2\pi$, and the correspondence

$$e^{i\theta} \to \begin{bmatrix} \cos \theta & -\sin \theta \\ \sin \theta & \cos \theta \end{bmatrix}$$

is an isomorphism from C to SO_2.

Now suppose that A belongs to SO_3. The characteristic polynomial $\det(A - \lambda I)$ is a cubic and therefore must have at least one real root. That is to say, A has a real eigenvalue. As the eigenvalues all have unit modulus, and as their product is $\det(A)$, we see that $+1$ is an eigenvalue of A. If $\mathbf{v}$ is a corresponding eigenvector, the line through the origin determined by $\mathbf{v}$ is left fixed by f_A. Also since f_A preserves right angles, it must send the plane which is perpendicular to $\mathbf{v}$, and which contains the origin, to itself. Construct an orthonormal basis for $\mathbb{R}^3$ which has the unit vector $\mathbf{v}/\|\mathbf{v}\|$ as first member. The matrix of f_A with respect to this new basis will be an element of SO_3 of the form

$$\begin{bmatrix} 1 & 0 & 0 \\ 0 & & \\ 0 & B & \end{bmatrix}.$$

Clearly $B \in SO_2$, so that f_A is a rotation with axis determined by $\mathbf{v}$. *Therefore, each matrix in SO_3 represents a rotation of $\mathbb{R}^3$ about an axis which passes through the origin.* Conversely, every rotation of $\mathbb{R}^3$ which fixes the origin is represented by a matrix in SO_3 (see Exercise 9.11).

If A lies in O_3 but not in SO_3, then $AU \in SO_3$ where

$$U = \begin{bmatrix} 1 & 0 & 0 \\ 0 & 1 & 0 \\ 0 & 0 & -1 \end{bmatrix}.$$

Note that U represents reflection in the (x, y) plane. We write A as the product $(AU)U$ to give

$$f_A = f_{AU} f_U.$$

As above f_{AU} is a rotation. Consequently, f_A is reflection in the (x, y) plane followed by a rotation.

We shall often refer to SO_3 as the *rotation group* in three dimensions. If a regular solid is positioned in $\mathbb{R}^3$ with its centre of gravity at the origin, then each of its symmetries is represented by a matrix in O_3. Its rotational symmetry group is therefore isomorphic to a subgoup of SO_3, and its full symmetry group to a subgroup of O_3.

EXAMPLE. The points $P = (1, 1, 1)$, $Q = (-1, -1, 1)$, $R = (1, -1, -1)$, and $S = (-1, 1, -1)$ are the vertices of a regular tetrahedron which has its centroid at the origin (see Figure 9.1). The two rotations about the axis through vertex P cyclically permute the coordinate axes and are represented by the matrices

$$\begin{bmatrix} 0 & 0 & 1 \\ 1 & 0 & 0 \\ 0 & 1 & 0 \end{bmatrix}, \quad \begin{bmatrix} 0 & 1 & 0 \\ 0 & 0 & 1 \\ 1 & 0 & 0 \end{bmatrix}.$$

The axis of symmetry which joins the midpoints of the edges PQ and RS is the z axis, and rotation through π about this axis has matrix

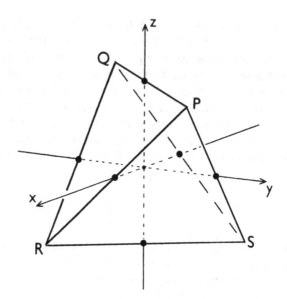

Figure 9.1

$$\begin{bmatrix} -1 & 0 & 0 \\ 0 & -1 & 0 \\ 0 & 0 & 1 \end{bmatrix}.$$

Reflection in the plane determined by P, Q, and O keeps P and Q fixed while interchanging R and S. The corresponding matrix is

$$\begin{bmatrix} 0 & 1 & 0 \\ 1 & 0 & 0 \\ 0 & 0 & 1 \end{bmatrix}.$$

The symmetry which cyclically permutes P, Q, R, S is neither a rotation nor a reflection, it is the product of three reflections. What is its matrix?

We shall classify finite subgroups of SO_3 in Chapter 19. Any such subgroup is either isomorphic to a cyclic group, isomorphic to a dihedral group, or isomorphic to the rotational symmetry group of a regular solid.

The matrices in $GL_n(\mathbb{C})$ correspond to linear transformations of the standard n-dimensional complex vector space $\mathbb{C}^n$. If $\mathbf{z} \in \mathbb{C}^n$, the length of $\mathbf{z}$ is the square root of $\mathbf{z}\mathbf{z}^*$, where * denotes complex conjugation. A complex matrix U is called a **unitary matrix** if $U^{*t}U$ is the identity matrix, and the unitary matrices are precisely those whose corresponding linear transformations preserve length in $\mathbb{C}^n$. The collection of all $n \times n$ unitary matrices forms a subgroup of $GL_n(\mathbb{C})$ called the **Unitary Group**, U_n. Those elements of U_n which have determinant equal to $+1$ form the **Special Unitary Group**, SU_n.

EXERCISES

In those exercises which involve the representation of linear transformations by matrices, we assume that the Euclidean space involved is equipped with the *standard* basis.

9.1. Which of the following collections of $n \times n$ real matrices form groups under matrix multiplication?

(a) The diagonal matrices whose diagonal entries are non-zero.
(b) The symmetric matrices.
(c) Those invertible matrices which have integer entries.
(d) Those invertible matrices which have rational entries.

9.2. Check that the set of all matrices of the form

$$\begin{bmatrix} a & b \\ 0 & c \end{bmatrix}, \qquad \text{where } a, b, c \in \mathbb{R} \text{ and } ac \neq 0$$

is a subgroup of $GL_2(\mathbb{R})$.

9.3. Prove that the elements of $GL_n(\mathbb{R})$ which have integer entries and determinant equal to $+1$ or -1 form a subgroup of $GL_n(\mathbb{R})$. This subgroup is denoted by $GL_n(\mathbb{Z})$.

9.4. The points $(1, -\sqrt{3})$, $(1, \sqrt{3})$, and $(-2, 0)$ determine an equilateral triangle. Work out the matrices in O_2 which represent plane symmetries of this triangle. Do the same calculation for the regular hexagon, whose corners have coordinates $(2, 0)$, $(1, \sqrt{3})$, $(-1, \sqrt{3})$, $(-2, 0)$, $(-1, -\sqrt{3})$, $(1, -\sqrt{3})$.

9.5. Let

$$A_\theta = \begin{bmatrix} \cos\theta & -\sin\theta \\ \sin\theta & \cos\theta \end{bmatrix} \quad \text{and} \quad B_\varphi = \begin{bmatrix} \cos\varphi & \sin\varphi \\ \sin\varphi & -\cos\varphi \end{bmatrix}.$$

Prove that $A_\theta A_\varphi = A_{\theta+\varphi}$, $A_\theta B_\varphi = B_{\theta+\varphi}$, $B_\theta A_\varphi = B_{\theta-\varphi}$, and $B_\theta B_\varphi = A_{\theta-\varphi}$, where the angles in the matrices are read mod 2π. Interpret these results geometrically.

9.6. Adopt the notation of the previous question and work out the products $A_\theta B_\varphi A_\theta^{-1}$, $B_\varphi A_\theta B_\varphi$, $A_\theta B_\varphi A_\theta^{-1} B_\varphi$. Evaluate each of these when $\theta = \frac{\pi}{3}$ and $\varphi = \frac{\pi}{4}$.

9.7. Complete the entries in

$$\begin{bmatrix} 1/\sqrt{2} & 0 & . \\ 0 & 1 & . \\ -1/\sqrt{2} & 0 & . \end{bmatrix}$$

to give an element of SO_3, and to give an element of $O_3 - SO_3$. Describe the linear transformations represented by these matrices.

9.8. Show that

$$\begin{bmatrix} 2/3 & 1/3 & 2/3 \\ -2/3 & 2/3 & 1/3 \\ -1/3 & -2/3 & 2/3 \end{bmatrix}, \quad \begin{bmatrix} -1/\sqrt{2} & 1/\sqrt{3} & 1/\sqrt{6} \\ 1/\sqrt{2} & 1/\sqrt{3} & 1/\sqrt{6} \\ 0 & 1/\sqrt{3} & -2/\sqrt{6} \end{bmatrix}$$

both represent rotations and find axes for these rotations.

9.9. Let v_1, v_2, v_3 be three mutually orthogonal vectors in $\mathbb{R}^3$ and let A be the matrix formed by taking v_1 as first column, v_2 as second column, and v_3 as third column. If

$$B = \begin{bmatrix} 1 & 0 & 0 \\ 0 & 1 & 0 \\ 0 & 0 & -1 \end{bmatrix}$$

show that ABA^{-1} represents reflection in the plane containing v_1 and v_2, and that $-ABA^{-1}$ represents rotation through π about the axis determined by v_3. Find the matrix which represents reflection in the plane $x + \sqrt{3}y = z$.

9.10. Prove that the correspondence

$$A \rightarrow \begin{cases} \begin{bmatrix} A & 0 \\ 0 & 1 \end{bmatrix} & \text{if } A \in SO_2 \\[2ex] \begin{bmatrix} A & 0 \\ 0 & -1 \end{bmatrix} & \text{if } A \in O_2 - SO_2 \end{cases}$$

is an isomorphism from O_2 to a subgroup of SO_3. Thicken the shapes of Exercise 9.4 to produce horizontal triangular and hexagonal plates, and write down the elements of SO_3 which represent the rotational symmetries of these plates.

9.11. Show that a rotation of $\mathbb{R}^3$ which fixes the origin is represented by a matrix in SO_3. First assume that the axis of the rotation is the z-axis, then deal with the general case.

9.12. Prove that the matrices

$$\begin{bmatrix} 1 & 0 & 0 \\ 0 & 1 & 0 \\ 0 & 0 & 1 \end{bmatrix}, \quad \begin{bmatrix} 1 & 0 & 0 \\ 0 & -1 & 0 \\ 0 & 0 & -1 \end{bmatrix},$$

$$\begin{bmatrix} -1 & 0 & 0 \\ 0 & 1 & 0 \\ 0 & 0 & -1 \end{bmatrix}, \quad \begin{bmatrix} -1 & 0 & 0 \\ 0 & -1 & 0 \\ 0 & 0 & 1 \end{bmatrix}$$

form a subgroup of SO_3 and find the corresponding rotations. Draw a picture of

$$\{(x, y, z) \in \mathbb{R}^3 | x^2 + (y - 3)^2 \leqslant 25, \ x^2 + (y + 3)^2 \leqslant 25, \ -1 \leqslant z \leqslant 1\}$$

and verify that it is a "*regular two-sided shape*" whose rotational symmetries are precisely those represented above. We often refer to the rotational symmetry group of this solid as the *dihedral group* D_2.

9.13. Check that the $n \times n$ unitary matrices do form a group under matrix multiplication, and that the determinant of each of these matrices is a complex number of modulus 1.

9.14. Show that the elements of U_2 have the form

$$\begin{bmatrix} z & w \\ -e^{i\theta} w* & e^{i\theta} z* \end{bmatrix}$$

where $z, w \in \mathbb{C}$, $\theta \in \mathbb{R}$ and $zz* + ww* = 1$. Which of these matrices belong to SU_2?

CHAPTER 10

Products

The **direct product** $G \times H$ of two groups G and H is constructed as follows. Its elements are ordered pairs (g, h) where $g \in G$ and $h \in H$, with multiplication defined by

$$(g, h)(g', h') = (gg', hh').$$

Both g and g' are elements of G, and they are multiplied in G to give the first entry, gg', of this product. The second entry is obtained by multiplying h and h' in H. Therefore (gg', hh') is an element of $G \times H$. Associativity follows directly from associativity in both G and H. The pair (e, e) is the identity, and (g^{-1}, h^{-1}) is the inverse of (g, h). So $G \times H$ is a group. (We hope our use of the same symbol for the identity elements of G and H will not cause confusion.) The correspondence $(g, h) \to (h, g)$ makes it clear that $G \times H$ is isomorphic to $H \times G$. If either G or H is an infinite group, then $G \times H$ is infinite, otherwise the order of $G \times H$ is the product of the orders of G and H. If G and H are both abelian, then $G \times H$ is abelian. Now G is isomorphic to the subgroup $\{(g, e) | g \in G\}$ of $G \times H$ via the correspondence $g \to (g, e)$, and H is isomorphic to the subgroup $\{(e, h) | h \in H\}$ via $h \to (e, h)$. Since any subgroup of an abelian group is abelian, we see that if $G \times H$ is abelian, then so are both G and H. The direct product $G_1 \times \cdots \times G_n$ of a finite collection of groups has elements $(x_1, \ldots, x_n)$ where $x_i \in G_i$, $1 \leqslant i \leqslant n$, which are combined via

$$(x_1, \ldots, x_n)(x'_1, \ldots, x'_n) = (x_1 x'_1, \ldots, x_n x'_n).$$

Again, changing the order of the factors always produces an isomorphic group.

EXAMPLES. (i) $\mathbb{Z}_2 \times \mathbb{Z}_3$ has six elements, $(0, 0)$, $(1, 0)$, $(0, 1)$, $(1, 1)$, $(0, 2)$, $(1, 2)$, which are combined by

$$(x, y) + (x', y') = (x +_2 x', y +_3 y').$$

It seems sensible to use $+$ for the group structure, since we have "addition" in each factor. We shall follow this convention whenever we have products of cyclic groups. By repeatedly adding the element $(1, 1)$ to itself we can fill out the whole group. Therefore, $\mathbb{Z}_2 \times \mathbb{Z}_3$ is *cyclic* and must be isomorphic to $\mathbb{Z}_6$. A specific isomorphism from $\mathbb{Z}_2 \times \mathbb{Z}_3$ to $\mathbb{Z}_6$ is given by

$$(0, 0) \to 0, \quad (1, 1) \to 1, \quad (0, 2) \to 2,$$
$$(1, 0) \to 3, \quad (0, 1) \to 4, \quad (1, 2) \to 5.$$

(ii) In a similar fashion we can write down the four elements of $\mathbb{Z}_2 \times \mathbb{Z}_2$ as $(0, 0)$, $(1, 0)$, $(0, 1)$, $(1, 1)$, this time taking addition modulo 2 in both coordinates. Each non-identity element now has order 2, so the group is not cyclic. It is isomorphic to the group of plane symmetries of a chessboard (Chapter 7) via the correspondence

$$(0, 0) \to e, \quad (1, 0) \to q_1,$$
$$(0, 1) \to q_2, \quad (1, 1) \to r.$$

$\mathbb{Z}_2 \times \mathbb{Z}_2$ is often called **Klein's group**.

(iii) We write $\mathbb{R}^n$ for the direct product of n copies of $\mathbb{R}$. In the usual way, we think of elements of $\mathbb{R}^n$ as vectors $\mathbf{x} = (x_1, \ldots, x_n)$, and the group operation is just vector addition written

$$\mathbf{x} + \mathbf{y} = (x_1 + y_1, \ldots, x_n + y_n).$$

(10.1) Theorem. $\mathbb{Z}_m \times \mathbb{Z}_n$ *is cyclic if and only if the highest common factor of m and n is* 1.

Proof. Let k be the order of the element $(1, 1)$ in $\mathbb{Z}_m \times \mathbb{Z}_n$. Adding $(1, 1)$ to itself k times gives $(0, 0)$, in other words

$$(k(\mathrm{mod}\, m), k(\mathrm{mod}\, n)) = (0, 0).$$

This means that m and n are both factors of k. If the highest common factor of m and n is 1, then mn must be a factor of k, and therefore $k = mn$. So in this case $(1, 1)$ generates $\mathbb{Z}_m \times \mathbb{Z}_n$ and we have a cyclic group.

Now let d be the highest common factor of m and n, and suppose d is greater than 1. We must show that $\mathbb{Z}_m \times \mathbb{Z}_n$ is not cyclic. Let $m' = m/d$ and $n' = n/d$. For any element (x, y) of $\mathbb{Z}_m \times \mathbb{Z}_n$, we have

$$m'dn'(x, y) = (m'dn'x(\mathrm{mod}\, m), m'dn'y(\mathrm{mod}\, n))$$
$$= (mn'x(\mathrm{mod}\, m), m'ny(\mathrm{mod}\, n))$$
$$= (0, 0),$$

so the order of (x, y) is at most $m'dn'$. Therefore, $\mathbb{Z}_m \times \mathbb{Z}_n$ does not contain an element of order mn and consequently cannot be cyclic. $\qquad \square$

EXAMPLE (iv). Let I denote the 3×3 identity matrix and write J for $-I$. Both I and J commute with every other matrix in O_3, and together they form a subgroup of O_3 of order 2. We shall show that O_3 is isomorphic to the direct product of SO_3 and this subgroup. Define

$$\varphi: SO_3 \times \{I, J\} \to O_3$$

by $\varphi(A, U) = AU$, where $A \in SO_3$ and $U \in \{I, J\}$. Then φ preserves the algebraic structures involved because

$$\varphi((A, U)(B, V)) = \varphi(AB, UV)$$

$$= ABUV$$

$$= AUBV$$

$$= \varphi(A, U)\varphi(B, V)$$

for all $A, B \in SO_3$ and $U, V \in \{I, J\}$. If $\varphi(A, U) = \varphi(B, V)$, then $AU = BV$, giving $\det(AU) = \det(BV)$. But

$$\det(AU) = \det(A) . \det(U) = \det(U)$$

because $A \in SO_3$, and similarly $\det(BV) = \det(V)$. Hence, $U = V$, $A = B$, and we conclude that φ is injective. It only remains to check that φ is surjective. Given $A \in O_3$, either $A \in SO_3$, in which case $A = \varphi(A, I)$, or $AJ \in SO_3$ and $A = \varphi(AJ, J)$. This completes the argument.

We note that $\{I, J\}$ is isomorphic to $\mathbb{Z}_2$; just send I to 0 and J to 1. Therefore, O_3 is isomorphic to $SO_3 \times \mathbb{Z}_2$. The same argument shows that O_n is isomorphic to $SO_n \times \mathbb{Z}_2$ when n is odd. For even n this result is false (see Exercise 10.9).

The above example generalises as follows: If H and K are subsets of a group G, let HK denote the collection of all products xy where $x \in H$, $y \in K$.

(10.2) Theorem. *If H and K are subgroups of G for which $HK = G$, if they have only the identity element in common, and if every element of H commutes with every element of K, then G is isomorphic to $H \times K$.*

Proof. Mimic the argument of Example (iv). Define

$$\varphi: H \times K \to G$$

by $\varphi(x, y) = xy$ for all $x \in H$, $y \in K$. Then

$$\varphi((x, y)(x', y')) = \varphi(xx', yy')$$

$$= xx'yy'$$

$$= xyx'y', \qquad \text{because elements of } H \text{ commute with elements of } K$$

$$= \varphi(x, y)\varphi(x', y').$$

So φ takes the multiplication of $H \times K$ to that of G. If $\varphi(x, y) = \varphi(x', y')$, then $xy = x'y'$ and, therefore,

$$(x')^{-1}x = y'y^{-1}.$$

As the left-hand side belongs to H and the right-hand side to K, both belong to $H \cap K$ and must therefore be the identity. Thus, $x = x'$, $y = y'$, and φ is injective. We also know that $HK = G$, which means that every element of G can be expressed as a product xy for some $x \in H$, $y \in K$. Therefore, φ is surjective and provides us with an isomorphism from $H \times K$ to G. $\square$

The linear transformation $f_j: \mathbb{R}^3 \to \mathbb{R}^3$ sends each vector $\mathbf{x}$ to $-\mathbf{x}$ and is called **central inversion**. Place a regular solid in $\mathbb{R}^3$ with its centre of gravity at the origin. Then, with the exception of the tetrahedron, it has central inversion as one of its symmetries. If G is the full symmetry group of the solid, and if H is the subgroup of rotational symmetries (so elements of H correspond to matrices in SO_3), an application of Theorem (10.2) shows that G is isomorphic to $H \times \langle f_j \rangle$ and thus to $H \times \mathbb{Z}_2$. Therefore, *the full symmetry groups of the cube and octahedron are isomorphic to* $S_4 \times \mathbb{Z}_2$, *and those of the dodecahedron and icosahedron are isomorphic to* $A_5 \times \mathbb{Z}_2$.

EXERCISES

10.1. If $G \times H$ is cyclic, prove that G and H are both cyclic.

10.2. Show that $\mathbb{Z} \times \mathbb{Z}$ is not isomorphic to $\mathbb{Z}$.

10.3. Prove that $\mathbb{C}$ is isomorphic to $\mathbb{R} \times \mathbb{R}$, and that $\mathbb{C} - \{0\}$ is isomorphic to $\mathbb{R}^{pos} \times C$.

10.4. Klein's group is often referred to as the *four group* (Vierergruppe) and denoted by V. Show that $\mathbb{Z}_3 \times V$ is isomorphic to $\mathbb{Z}_2 \times \mathbb{Z}_6$.

10.5. Show that the "diagonal" $\{(x, x) | x \in G\}$ is a subgroup of $G \times G$, and that this subgroup is isomorphic to G.

10.6. If A is a subgroup of G, and if B is a subgroup of H, check that $A \times B$ is a subgroup of $G \times H$. Find a subgroup of $\mathbb{Z} \times \mathbb{Z}$ which does not occur in this way.

10.7. Which of the following groups are isomorphic to one another?

$$\mathbb{Z}_{24}, \qquad D_4 \times \mathbb{Z}_3, \qquad D_{12}, \qquad A_4 \times \mathbb{Z}_2,$$

$$\mathbb{Z}_2 \times D_6, \qquad S_4, \qquad \mathbb{Z}_{12} \times \mathbb{Z}_2.$$

10.8. The element $(\varepsilon, 1)$ of $A_n \times \mathbb{Z}_2$ commutes with every element of $A_n \times \mathbb{Z}_2$. Use this observation to prove that $A_n \times \mathbb{Z}_2$ is not isomorphic to S_n when $n \geqslant 3$.

10.9. Why does the construction of Example (iv) not lead to an isomorphism from $SO_n \times \mathbb{Z}_2$ to O_n when n is even? Which elements of O_n commute with every element of O_n? Show that $SO_n \times \mathbb{Z}_2$ is not isomorphic to O_n when n is even.

10.10. Let G be the group whose elements are infinite sequences $(a_1, a_2, \ldots)$ of integers which combine termwise via

$$(a_1, a_2, \ldots)(b_1, b_2, \ldots) = (a_1 + a_2, b_1 + b_2, \ldots).$$

Prove that $G \times \mathbb{Z}$ and $G \times G$ are both isomorphic to G.

10.11. Show that D_{2n} is isomorphic to $D_n \times \mathbb{Z}_2$ when n is odd.

10.12. If G is a group of order 4 which is not cyclic, prove that G is isomorphic to Klein's group.

10.13. Let G be a finite group in which every element other than the identity has order 2. Prove that G is isomorphic to the direct product of a finite number of copies of $\mathbb{Z}_2$.

CHAPTER 11

Lagrange's Theorem

Consider a finite group G together with a subgroup H of G. *Are the orders of H and G related in any way?* Assuming H is not all of G, choose an element g_1 from $G - H$, and multiply every element of H on the left by g_1 to form the set

$$g_1 H = \{g_1 h | h \in H\}.$$

We claim that $g_1 H$ has the same size as H and is disjoint from H. The first assertion follows because the correspondence $h \to g_1 h$ from H to $g_1 H$ can be inverted (just multiply every element of $g_1 H$ on the left by g_1^{-1}) and is therefore a bijection. For the second, suppose x lies in both H and $g_1 H$. Then there is an element $h_1 \in H$ such that $x = g_1 h_1$. But this gives $g_1 = x h_1^{-1}$, which contradicts our initial choice of g_1 *outside* H.

If H and $g_1 H$ together fill out all of G, then clearly $|G| = 2|H|$. Otherwise we choose $g_2 \in G - (H \cup g_1 H)$ and form $g_2 H$. Again, this has the same number of elements as H and is disjoint from H. We hope that it does not meet $g_1 H$. To check this, suppose x lies in $g_1 H$ and $g_2 H$. Then there are elements h_1, h_2 of H such that $x = g_1 h_1 = g_2 h_2$, giving $g_2 = g_1 (h_1 h_2^{-1})$ and contradicting our choice of g_2 outside $g_1 H$. If H, $g_1 H$, and $g_2 H$ together fill out G, then $|G| = 3|H|$. If not, we choose g_3 in their complement and continue, checking that $g_3 H$ does not meet any of H, $g_1 H$, or $g_2 H$. As G is finite, this process stops after a finite number of steps, and if there are k steps we find G broken up as the union of $k + 1$ pieces

$$H, g_1 H, \ldots, g_k H$$

no two of which overlap, and each of which has the same size as H. Consequently, $|G| = (k + 1)|H|$. We have proved the following result:

(11.1) Lagrange's Theorem. *The order of a subgroup of a finite group is always a divisor of the order of the group.*

A word of caution is appropriate at this point. If G is a finite group, and if m is a divisor of the order of G, there is no guarantee that G will contain a subgroup of order m. Indeed, A_4 has no subgroup of order 6, as we shall see later. This is disappointing, but we can salvage something. In Chapter 13 we shall prove that if p is a *prime* divisor of the order of G, then G does contain a subgroup of order p, and a second existence theorem for subgroups can be found in the chapter dealing with Sylow's theorems.

It is instructive to carry out the above procedure for a particular example. Suppose G is S_3 and the subgroup is $H = \{\varepsilon, (13)\}$. Select g_1 outside H; say $g_1 = (123)$, then

$$g_1 H = \{(123)\varepsilon, (123)(13)\} = \{(123), (23)\}.$$

Select g_2 outside $H \cup g_1 H$; say $g_2 = (12)$, then

$$g_2 H = \{(12)\varepsilon, (12)(13)\} = \{(12), (132)\}.$$

The group is now broken up into three disjoint subsets, $H, g_1 H, g_2 H$, each of which contains two elements.

Here are some useful corollaries of Lagrange's theorem; G denotes a finite group throughout.

(11.2) Corollary. *The order of every element of G is a divisor of the order of G.*

Proof. Remember that the order of an element is equal to the order of the subgroup generated by that element. $\square$

(11.3) Corollary. *If G has prime order, then G is cyclic.*

Proof. If $x \in G - \{e\}$, the order of x must equal the order of G by (11.2). Therefore, $\langle x \rangle = G$. $\square$

(11.4) Corollary. *If x is an element of G then $x^{|G|} = e$.*

Proof. Let m be the order of x. By (11.2) we know that $|G| = km$ for some integer k. Hence

$$x^{|G|} = x^{km} = (x^m)^k = e. \qquad \square$$

Let n be a positive integer and consider the collection R_n of all those integers m which satisfy $1 \leqslant m \leqslant n - 1$, and for which the highest common factor of m and n is 1. We claim that *multiplication modulo n makes R_n into a group.* (This answers the question raised at the end of Chapter 3.) If m_1 and m_2 belong to R_n, the highest common factor of $m_1 m_2$ and n is certainly 1. Therefore, $m_1 m_2 \pmod{n}$ and n have highest common factor 1, verifying that R_n is closed under multiplication modulo n. Associativity is easily checked (Exercise 3.5) and the integer 1 acts as identity. Finally, if $m \in R_n$ there are integers x and y

such that $xm + yn = 1$. Reading this equation mod n provides a multiplicative inverse for m, namely $x(\mathrm{mod}\, n)$. Clearly R_n is abelian. Its order is written $\varphi(n)$ and φ is called Euler's phi-function.

EXAMPLES. (i) The elements of R_9 are 1, 2, 4, 5, 7, 8 and $\varphi(9) = 6$. As $2^2 = 4$, $2^3 = 8$, $2^4 = 16(\mathrm{mod}\, 9) = 7$, and $2^5 = 32(\mathrm{mod}\, 9) = 5$, we see that R_9 is cyclic generated by the integer 2.

(ii) The elements of R_{16} are 1, 3, 5, 7, 9, 11, 13, 15 and $\varphi(16) = 8$. Let $H = \langle 3 \rangle$ and $K = \langle 15 \rangle$. We easily check that $H = \{1, 3, 9, 11\}$, $K = \{1, 15\}$, $HK = R_{16}$ and $H \cap K = \{1\}$. By (10.2) we have

$$R_{16} \cong H \times K \cong \mathbb{Z}_4 \times \mathbb{Z}_2.$$

(11.5) Euler's Theorem. *If the highest common factor of x and n is 1, then $x^{\varphi(n)}$ is congruent to 1 modulo n.*

Proof. Divide x by n to give a remainder m which belongs to R_n. By (11.4) we know that $m^{\varphi(n)}$ is congruent to 1 modulo n. Since $x^{\varphi(n)}$ is congruent to $m^{\varphi(n)}$ modulo n, the result follows. □

(11.6) Fermat's Little Theorem. *If p is prime and if x is not a multiple of p, then x^{p-1} is congruent to 1 modulo p.*

Proof. Apply Euler's theorem noting that $\varphi(p) = p - 1$. □

By Lagrange's theorem, the order of a subgroup of A_4 must be a factor of 12. The extreme cases 1 and 12 correspond to the subgroup which consists just of the identity element and to the whole group, respectively. A subgroup of order 2 will contain the identity plus an element of order 2, so we have three possibilities $\{\varepsilon, (12)(34)\}$, $\{\varepsilon, (13)(24)\}$, and $\{\varepsilon, (14)(23)\}$. There are four subgroups of order 3, each generated by a 3-cycle, namely

$$\{\varepsilon, (123), (132)\}, \qquad \{\varepsilon, (124), (142)\},$$
$$\{\varepsilon, (134), (143)\}, \qquad \{\varepsilon, (234), (243)\}.$$

There is a single subgroup of order 4

$$\{\varepsilon, (12)(34), (13)(24), (14)(23)\}.$$

It is unique because all the other elements of A_4 are 3-cycles, and a 3-cycle cannot belong to a group of order 4 by (11.2).

As we mentioned earlier, A_4 does not contain a subgroup of order 6. For suppose H is a subgroup of A_4, which has six elements. If a 3-cycle belongs to H, its inverse must also belong to H, so the number of 3-cycles in H is *even*. There cannot be six as we need room for the identity element. Suppose there are four, say α, α^{-1}, β, and β^{-1}. Then ε, α, α^{-1}, β, β^{-1}, $\alpha\beta$, $\alpha\beta^{-1}$ are all distinct and belong to H, contradicting our assumption that $|H| = 6$. Finally, if only two 3-cycles lie in H, then H must contain the subgroup $\{\varepsilon, (12)(34), (13)(24), (14)(23)\}$. But 4 is not a factor of 6, so Lagrange's theorem rules out this case too. We conclude that no such subgroup of order 6 exists.

EXERCISES

11.1. Carry out the procedure used in the proof of Lagrange's theorem, taking $G = D_6$, $H = \langle r \rangle$, then $G = D_6$, $H = \langle r^3 \rangle$, and finally $G = A_4$, $H = \langle (234) \rangle$.

11.2. Let H be a subgroup of a group G. Prove that $g_1 H = g_2 H$ if and only if $g_1^{-1} g_2$ belongs to H.

11.3. If H and K are finite subgroups of a group G, and if their orders are relatively prime, show that they have only the identity element in common.

11.4. Suppose the order of G is the product of two distinct primes. Show that any proper subgroup of G must be cyclic.

11.5. Given subsets X and Y of a group G, write XY for the set of all products xy where $x \in X$ and $y \in Y$. If X and Y are both finite, if Y is a subgroup of G, and if XY is contained in X, prove that the size of X is a multiple of the size of Y.

11.6. If the highest common factor of m and n is 1, show that R_{mn} is isomorphic to the product group $R_m \times R_n$. Use this observation to check that R_{20} is isomorphic to $\mathbb{Z}_2 \times \mathbb{Z}_4$.

11.7. Let n be a positive integer and let m be a factor of $2n$. Show that D_n contains a subgroup of order m.

11.8. Does A_5 contain a subgroup of order m for each factor m of 60?

11.9. Let G be a finite abelian group and let m be the least common multiple of the orders of its elements. Prove that G contains an element of order m.

11.10. Supply a finite non-abelian group for which the conclusion of the previous exercise fails.

11.11. If H is a subgroup of a finite group G, and if $|G| = m|H|$, adapt the proof of Lagrange's theorem to show that $g^{m!} \in H$ for all $g \in G$.

11.12. Prove that R_p is a cyclic group when p is a prime number.

CHAPTER 12

Partitions

A *partition* of a set X is a decomposition of the set into non-empty subsets, no two of which overlap and whose union is all of X. The proof of Lagrange's theorem involved partitioning a group into subsets, each of which had the same number of elements as a given subgroup. In this chapter we shall show how to *recognise* partitions.

Suppose we have a partition of X, and let x, y be points of X. We shall say that x *is related to* y if x lies in the same member of the partition as y. The following properties are immediate:

(a) Each $x \in X$ is related to itself.
(b) If x is related to y, then y is related to x, for any two points $x, y \in X$.
(c) If x is related to y and if y is related to z, then x is related to z, for any three points $x, y, z \in X$.

This may seem rather abstract, but Figure 12.1 should help. The shaded areas represent the different members of the partition.

We now change our point of view. Let X be a set and let $\mathcal{R}$ be a subset of the cartesian product $X \times X$. In other words, $\mathcal{R}$ is a collection of ordered pairs (x, y) whose coordinates x, y come from X. *Given two points x and y of X we shall say that x is related to y if the ordered pair (x, y) happens to lie in $\mathcal{R}$.* If properties (a), (b), and (c) are valid, then we call $\mathcal{R}$ an *equivalence relation* on X. For each $x \in X$ the collection of all points which are related to it is written $\mathcal{R}(x)$ and called the *equivalence class* of x.

(12.1) Theorem. $\mathcal{R}(x) = \mathcal{R}(y)$ *whenever* $(x, y) \in \mathcal{R}$.

Proof. Suppose $(x, y) \in \mathcal{R}$ and let $z \in \mathcal{R}(x)$. Then z is related to x. But x is related to y, so by property (c) we know that z is related to y. Hence, $z \in \mathcal{R}(y)$

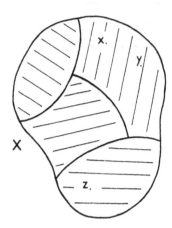

x is related to y

x is not related to z

Figure 12.1

and we have $\mathcal{R}(x) \subseteq \mathcal{R}(y)$. We also know that $(y, x) \in \mathcal{R}$ by property (b), so reversing the roles of x and y gives us $\mathcal{R}(y) \subseteq \mathcal{R}(x)$ and completes the argument.   □

EXAMPLE (i). Let $X = \mathbb{Z}$ and let $\mathcal{R}$ consist of all ordered pairs (x, y) in $\mathbb{Z} \times \mathbb{Z}$ for which $x - y$ is divisible by 3. Certainly $x - x$ is divisible by 3 for any integer x. If $x - y$ is divisible by 3, then so is $y - x$, and if both $x - y$ and $y - z$ are divisible by 3, then so is their sum $x - z$. So $\mathcal{R}$ is an equivalence relation on $\mathbb{Z}$. Any integer x is related to either 0, 1, or 2; consequently, there are three distinct equivalence classes. The equivalence class of 0 consists of all multiples of 3, that of 1 contains all integers which are congruent to 1 modulo 3, and that of 2 the remaining integers which are congruent to 2 modulo 3. Notice that $\mathcal{R}(0)$, $\mathcal{R}(1)$, and $\mathcal{R}(2)$ form a *partition* of $\mathbb{Z}$ (see Fig. 12.2). This gives a clue to the statement of our next result.

(12.2) Theorem. *The distinct equivalence classes of an equivalence relation on X form a partition of X.*

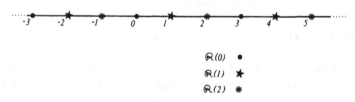

Figure 12.2

Proof. Each equivalence class is non-empty because $\mathscr{R}(x)$ always contains x by property (a). If $\mathscr{R}(x)$ and $\mathscr{R}(y)$ overlap, there must be a point z in their intersection. Then z is related to both x and y. By property (b), x is related to z, and therefore also related to y by property (c). We conclude that $\mathscr{R}(x) = \mathscr{R}(y)$. So two equivalence classes can only overlap if they coincide. Finally, as each point x lies in its own equivalence class $\mathscr{R}(x)$, the union of the equivalence classes is all of X. $\qquad\square$

EXAMPLE (ii). Replacing 3 in the previous example by the positive integer n gives an equivalence relation on $\mathbb{Z}$ which partitions $\mathbb{Z}$ into n equivalence classes $\mathscr{R}(0)$, $\mathscr{R}(1)$, $\ldots$, $\mathscr{R}(n-1)$ called *congruence classes*. An integer x belongs to $\mathscr{R}(m)$ precisely when x is congruent to m modulo n.

EXAMPLE (iii). Let H be a subgroup of G and let $\mathscr{R}$ be the collection of ordered pairs (x,y) with entries from G for which $y^{-1}x \in H$. It is easy to check that $\mathscr{R}$ is an equivalence relation on G. (For any $x \in G$ we have $x^{-1}x = e \in H$, if $y^{-1}x \in H$, then $x^{-1}y = (y^{-1}x)^{-1} \in H$, and if $y^{-1}x$, $z^{-1}y \in H$, then $z^{-1}x = (z^{-1}y)(y^{-1}x) \in H$.) The equivalence class of a particular element $g \in G$ consists of all those $x \in G$ which satisfy $g^{-1}x \in H$. Now $g^{-1}x$ belongs to H precisely when $x = gh$ for some element $h \in H$. Therefore, $\mathscr{R}(g)$ is the set gH obtained by multiplying every element of H on the left by g. This set gH is called the *left coset* of H determined by g. By (12.2) we know that the distinct left cosets of H in G form a partition of G, precisely what was needed in the proof of Lagrange's theorem. If $\mathscr{R}$ is changed to the collection of ordered pairs $(x, y) \in G \times G$, for which $xy^{-1} \in H$, we again have an equivalence relation on G. This time the equivalence class of g is the *right coset* Hg, obtained if we multiply every element of H on the right by g.

Nothing very new has emerged so far. We managed a perfectly satisfactory proof of Lagrange's theorem without (12.2), just using our common sense. The strength of (12.2) is its *generality*. It will allow us to check quickly and easily that certain decompositions of sets are partitions. We hint at two applications below. Both will be important in later chapters.

EXAMPLE (iv). Let x,y be elements of a group G. We say that x is *conjugate* to y if $gxg^{-1} = y$ for some $g \in G$. The collection of all elements conjugate to a given element is called a *conjugacy class*, and we claim that the distinct conjugacy classes form a partition of G. Let $\mathscr{R}$ be the subset of $G \times G$ consisting of pairs (x, y) for which x is conjugate to y. Each $x \in G$ is conjugate to itself because $exe^{-1} = x$. If x is conjugate to y, say $gxg^{-1} = y$, then y is conjugate to x because $g^{-1}yg = x$. Finally, if x is conjugate to y and y to z, say $g_1xg_1^{-1} = y$, $g_2yg_2^{-1} = z$, then x is conjugate to z because

$$(g_2g_1)x(g_2g_1)^{-1} = g_2(g_1xg_1^{-1})g_2^{-1}$$

$$= g_2yg_2^{-1}$$

$$= z.$$

Therefore, $\mathscr{R}$ is an equivalence relation on G and its equivalence classes, the conjugacy classes mentioned above, partition G. These conjugacy classes will be worked out for various groups in Chapter 14.

EXAMPLE (v). Let X be a set and let G be a subgroup of S_X, so that each element of G permutes the points of X. Let $\mathscr{R}$ be the subset of $X \times X$ defined as follows. An ordered pair (x, y) lies in $\mathscr{R}$ if $g(x) = y$ for some $g \in G$. This is an equivalence relation on X whose equivalence classes are called the orbits of G. (As usual the verification is easy. For each $x \in X$ we have $\varepsilon(x) = x$, so x is related to itself. If x is related to y, say $g(x) = y$, then $g^{-1}(y) = x$ which shows that y is related to x. Finally, if x is related to y and y to z, say $g(x) = y$ and $g'(y) = z$, then $g'g(x) = g'(y) = z$, showing that x is related to z.) It will be important to us later that the distinct orbits always form a partition of X, and we can be sure of this by (12.2). To make this more concrete, take X to be $\mathbb{R}^3$ and G to be the group of linear transformations f_A for which $A \in SO_3$. The origin is fixed by every linear transformation, so the orbit of $\mathbf{0}$ is just $\{\mathbf{0}\}$. Remembering that orthogonal transformations preserve length, it is not hard to check that the orbit of a non-zero vector $\mathbf{x}$ is the sphere which has the origin as its centre and $\|\mathbf{x}\|$ as its radius. The spheres of different radii, together with the point at the origin, do indeed form a partition of $\mathbb{R}^3$.

We end this chapter by introducing a group whose elements are most naturally defined as the equivalence classes of an equivalence relation.

EXAMPLE (vi). Begin with a pair of horizontal planes, three points in the top plane and the corresponding three vertically below them in the lower plane. Add strings which join the top points to the lower ones (Fig. 12.3). These strings must not intersect, and an individual string should meet each level between our two planes exactly once. Such a configuration is called a **braid**. Given two braids b_1, b_2 we can "multiply them" to give a new braid $b_1 b_2$ by simply stacking b_2 on top of b_1 as in Figure 12.4.

We write e for the trivial braid whose strings are vertical. The essential quality of a braid is the way in which its strings wind around one another, so the trivial braid seems to act as an identity for our multiplication (Fig. 12.5). Reflecting a braid b in the lower plane gives a braid which we denote by b^{-1}. Notice that $b^{-1}b$ is to all intents and purposes trivial (Fig. 12.6). We are very close to the structure of a group, and the notion of an equivalence relation is ideally suited to handle the remaining lack of precision.

If b_1 and b_2 are braids, we shall say that b_1 is related to b_2 provided the strings of b_1 can be deformed in a continuous fashion until they land on top of those of b_2. During the deformation the strings must stay between the two horizontal planes, their end points should remain fixed, and they are not allowed to intersect. The resulting relation $\mathscr{R}$ is an equivalence relation on the collection of all braids, and its equivalence classes form a group under the multiplication

$$\mathscr{R}(b_1)\mathscr{R}(b_2) = \mathscr{R}(b_1 b_2),$$

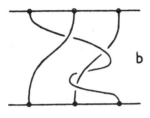

Figure 12.3

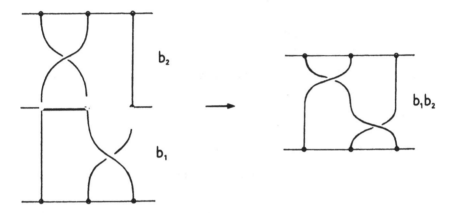

Figure 12.4

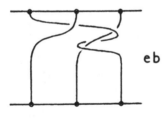

Figure 12.5

Figure 12.6

as the reader may easily check. It should not cause any surprise when we say that $\mathcal{R}(e)$ is the identity element, and that $\mathcal{R}(b^{-1})$ is the inverse of $\mathcal{R}(b)$. This is the **braid group** B_3 *on three strings*. There is of course a corresponding group B_n for each positive integer n.

EXERCISES

12.1. Which of the following subsets of $\mathbb{R} \times \mathbb{R}$ are equivalence relations on $\mathbb{R}$?

(a) $\{(x, y)|x - y$ is an even integer$\}$
(b) $\{(x, y)|x - y$ is rational$\}$
(c) $\{(x, y)|x + y$ is rational$\}$
(d) $\{(x, y)|x - y \geqslant 0\}$.

12.2. Do any of the following define equivalence relations on the set of all non-zero complex numbers?

(a) z is related to w if zw is real.
(b) z is related to w if z/w is real.
(c) z is related to w if z/w is an integer.

12.3. Find a group G and a subgroup H for which $\{(x, y)|xy \in H\}$ is not an equivalence relation on G.

12.4. Supply a group G and a subgroup H for which $\{(x, y)|xyx^{-1}y^{-1} \in H\}$ is not an equivalence relation on G.

12.5. Let $\mathcal{R}$ be an equivalence relation on X. Given $x \in X$, choose $y \in X$ such that $(x, y) \in \mathcal{R}$. Property (b) gives $(y, x) \in \mathcal{R}$, and property (c) then shows that $(x, x) \in \mathcal{R}$. Therefore, the first property of an equivalence relation seems to be redundant. What is wrong with this argument?

12.6. Let n be a positive integer and consider the equivalence relation on $\mathbb{Z}$ given by congruence modulo n. Write $[x]$ for the congruence class of x and define the *sum* of two such classes by

$$[x] + [y] = [x + y].$$

At first sight this rule seems to depend on the particular representatives chosen from the two equivalence classes. Show that addition is in fact *well defined*; in other words, if $[x] = [x']$ and $[y] = [y']$, then $[x + y] = [x' + y']$. Prove that the collection of congruence classes forms an abelian group under this operation, and that the resulting group is isomorphic to $\mathbb{Z}_n$. (Indeed, many authors prefer to define $\mathbb{Z}_n$ in this way.)

12.7. Work out the left and right cosets of H in G when

$$G = A_4, \qquad H = \{\varepsilon, (12)(34), (13)(24), (14)(23)\}$$

and when

$$G = A_4, \qquad H = \{\varepsilon, (123), (132)\}.$$

Verify that the left and right cosets coincide in the first case, but not in the second.

12.8. Let G be a finite group and let H be a subgroup of G which contains precisely half the elements of G. Show that $gH = Hg$ for every element g of G.

12.9. Here is a prescription which enables us to construct the *rational numbers* from the integers. Begin with the set X of all ordered pairs (m, n) of integers in which the second coordinate is non-zero. We think of (m, n) as representing the fraction m/n. Of course many different ordered pairs represent the same rational number, for example $(2, 3)$, $(4, 6)$, and $(-6, -9)$ all represent $\frac{2}{3}$. Somehow we have to identify all these pairs, and the notion of an equivalence relation is the appropriate tool. Agree that (m, n) is related to (m', n') whenever $mn' = m'n$. Show this is an equivalence relation on X. Write $[(m, n)]$ for the equivalence class of the ordered pair (m, n) and, motivated by the rules for adding and multiplying fractions, define

$$[(m, n)] + [(m_1, n_1)] = [(mn_1 + m_1 n, nn_1)],$$

$$[(m, n)] \cdot [(m_1, n_1)] = [(mm_1, nn_1)].$$

Check that both operations are well defined, that the set of all equivalence classes forms an abelian group under addition, and that if we remove the "zero class" $[(0, n)]$ the remainder form an abelian group under multiplication. Each rational number is represented by precisely one equivalence class and we have now modelled the algebraic structure of the rationals.

12.10. Consider the relation of conjugacy defined in Example (iv) and work out the conjugacy classes when $G = D_4$. Try to do the same for the infinite dihedral group D_∞.

12.11. Convince yourself that the braid group B_3 is infinite, is not abelian, and is generated by the two braids b_1, b_2 shown in Figure 12.4.

12.12. The construction of B_3 involves a pair of horizontal planes, three points in the top plane and the corresponding three vertically below them in the lower plane. Label the upper points, and the lower points, 1, 2, 3 so that points which are vertically aligned have the same label. Sliding along the strings of a braid now produces an element of S_3. Show that the function from B_3 to S_3 constructed in this way is surjective and sends the multiplication of B_3 to the multiplication of S_3. Find two different braids which both map to the permutation (123).

CHAPTER 13

Cauchy's Theorem

Here is the partial converse to Lagrange's theorem promised in Chapter 11.

(13.1) Cauchy's Theorem. *If p is a prime divisor of the order of a finite group G, then G contains an element of order p.*

Proof. We need an element $x \in G - \{e\}$ such that $x^p = e$. Consider the set X of all ordered strings $\mathbf{x} = (x_1, x_2, \ldots, x_p)$ of elements of G for which

$$x_1 x_2 \ldots x_p = e.$$

Our problem is to find such a string which has all its coordinates equal, but which is not $(e, e, \ldots, e)$. We shall do this by a careful analysis of the *size* of X.

How big is X? If the string $(x_1, x_2, \ldots, x_p)$ is to lie in X we may choose x_1, $x_2, \ldots, x_{p-1}$ arbitrarily from G, when x_p is completely determined by $x_p = (x_1 x_2 \ldots x_{p-1})^{-1}$. So the number of strings in X is $|G|^{p-1}$, which is a *multiple of p*.

Let $\mathscr{R}$ be the subset of $X \times X$ defined as follows. An ordered pair $(\mathbf{x}, \mathbf{y})$ belongs to $\mathscr{R}$ if $\mathbf{y}$ can be obtained by cyclically permuting the coordinates of $\mathbf{x}$. In other words $\mathbf{y}$ is one of

$$(x_1, x_2, \ldots, x_p)$$

$$(x_p, x_1, \ldots, x_{p-1})$$

$$\vdots$$

$$(x_2, \ldots, x_p, x_1).$$

$$(*)$$

Note that these cyclic permutations do all belong to X. For example,

$$x_p x_1 \ldots x_{p-1} = x_p(x_1 \ldots x_{p-1} x_p) x_p^{-1}$$
$$= x_p e x_p^{-1}$$
$$= e$$

which shows that $(x_p, x_1, \ldots, x_{p-1}) \in X$, and repeating this process deals with the others. One easily checks that $\mathscr{R}$ is an equivalence relation on X and that the equivalence class $\mathscr{R}(\mathbf{x})$ of the string $\mathbf{x} = (x_1, x_2, \ldots, x_p)$ is precisely the collection (∗).

Does cyclic permutation of the coordinates of a string always produce p *different* strings? Certainly not in the case of $\mathbf{e} = (e, e, \ldots, e)$ where cyclically permuting the entries does not give anything new, and $\mathscr{R}(\mathbf{e})$ contains *just one* element. The distinct equivalence classes of $\mathscr{R}$ partition X, so adding together the sizes of these classes gives the total number of elements in X. If every equivalence class other than $\mathscr{R}(\mathbf{e})$ contains p elements, then the size of X will be congruent to 1 modulo p, contradicting our earlier calculation. Therefore, there must be a string $\mathbf{x} = (x_1, x_2, \ldots, x_p)$, other than $\mathbf{e}$, whose equivalence class contains less than p elements. So two of the cyclic permutations in (∗) are equal, say

$$(x_{r+1}, \ldots, x_p, x_1, \ldots, x_r) = (x_{s+1}, \ldots, x_p, x_1, \ldots, x_s).$$

Assume $r > s$ and cycle back $p - r$ times to give

$$(x_1, x_2, \ldots, x_p) = (x_{k+1}, \ldots, x_p, x_1, \ldots, x_k)$$

where $k = p - r + s$. Equating corresponding coordinates we observe that $x_i = x_{k+i(\text{mod } p)}$ for $1 \leqslant i \leqslant p$, and consequently

$$x_1 = x_{k+1} = x_{2k+1} = \cdots = x_{(p-1)k+1}$$

where the suffices are read mod p. Suppose $ak + 1$ and $bk + 1$ are congruent modulo p where $0 \leqslant a < b \leqslant p - 1$. Then p divides $(b - a)k$, which is impossible because p is prime and both $b - a$ and k are less than p. Therefore the numbers

$$1, k + 1, 2k + 1, \ldots, (p - 1)k + 1$$

are all different when read mod p. As there are p of them, reading them mod p just gives $1, 2, \ldots, p$ possibly jumbled up in some different order. We conclude that $x_1 = x_2 = \cdots = x_p$, which gives $x_1^p = e$ as required. (Once we have a little more machinery at our disposal we will be able to steamline this type of argument; see Chapter 17.) □

As an application of Cauchy's theorem we show that a group of order 6 must be either cyclic or dihedral. We should be more precise.

(13.2) Theorem. *A group of order 6 is either isomorphic to $\mathbb{Z}_6$ or isomorphic to D_3.*

Proof. Let G be a group which contains six elements. Use Cauchy's theorem to select an element x of order 3 and an element y of order 2. The right cosets $\langle x \rangle$, $\langle x \rangle y$ give six elements $e, x, x^2, y, xy, x^2 y$ which fill out G. Now yx is one of these six and it is certainly not in $\langle x \rangle$ or equal to y. If $yx = xy$, then (10.2) shows that G is isomorphic to $\langle x \rangle \times \langle y \rangle$, and hence to $\mathbb{Z}_3 \times \mathbb{Z}_2$, which is cyclic by (10.1). Otherwise $yx = x^2 y$ and (in the notation of Section 4) changing x to r and y to s gives an isomorphism from G to D_3. $\square$

It is not much harder to show that if p is an odd prime, then any group of order $2p$ is either cyclic or dihedral (see Chapter 15).

We now have a good deal of information about groups of small order. Any group of order 2, 3, 5, or 7 is cyclic by (11.3), a group of order 4 is isomorphic to either $\mathbb{Z}_4$ or Klein's group (see Exercise 10.12), and any group of order 6 is cyclic or dihedral. The situation for order 8 is more complicated. We have already met four groups, each of which has eight elements, namely $\mathbb{Z}_8$, $\mathbb{Z}_4 \times \mathbb{Z}_2$, $\mathbb{Z}_2 \times \mathbb{Z}_2 \times \mathbb{Z}_2$, and D_4. Here is a fifth. A quaternion (or hypercomplex number) is an expression of the form $a + bi + cj + dk$, where a, b, c, d are real numbers and i, j, k satisfy

$$i^2 = j^2 = k^2 = -1, \qquad ij = -ji = k \qquad (*)$$

and the set of all quaternions is denoted by $\mathbb{H}$. The eight symbols $\pm 1, \pm i, \pm j, \pm k$ when multiplied according to $(*)$ form a group Q called the **quaternion group**. Its multiplication table is shown below

	1	-1	i	$-i$	j	$-j$	k	$-k$
1	1	-1	i	$-i$	j	$-j$	k	$-k$
-1	-1	1	$-i$	i	$-j$	j	$-k$	k
i	i	$-i$	-1	1	k	$-k$	$-j$	j
$-i$	$-i$	i	1	-1	$-k$	k	j	$-j$
j	j	$-j$	$-k$	k	-1	1	i	$-i$
$-j$	$-j$	j	k	$-k$	1	-1	$-i$	i
k	k	$-k$	j	$-j$	$-i$	i	-1	1
$-k$	$-k$	k	$-j$	j	i	$-i$	1	-1

Q is not abelian (so it cannot be isomorphic to one of $\mathbb{Z}_8$, $\mathbb{Z}_4 \times \mathbb{Z}_2$, $\mathbb{Z}_2 \times \mathbb{Z}_2 \times \mathbb{Z}_2$) and as -1 is the only element of order 2 it is not isomorphic to D_4, which contains five elements of order 2.

(13.3) Theorem. *A group of order 8 is isomorphic to one of the following*: $\mathbb{Z}_8$, $\mathbb{Z}_4 \times \mathbb{Z}_2$, $\mathbb{Z}_2 \times \mathbb{Z}_2 \times \mathbb{Z}_2$, D_4, Q.

Proof. Let G be a group which has eight elements. If there is an element of order 8, then G is isomorphic to $\mathbb{Z}_8$. *Suppose now that the largest order of an element of G is* 4. Choose an element x whose order is 4 and an element y from $G - \langle x \rangle$. The cosets $\langle x \rangle$, $\langle x \rangle y$ fill out G and provide the elements

$$e, x, x^2, x^3, y, xy, x^2y, x^3y.$$

We know that yx is not in $\langle x \rangle$, cannot equal y (as $yx = y$ gives $x = e$) and cannot equal x^2y (because $yx = x^2y$ leads to $x = y^{-1}x^2y$, which in turn gives $x^2 = y^{-1}x^2yy^{-1}x^2y = e$). Therefore, yx is either xy or x^3y. In addition the order of the element y is 2 or 4. Observe that y^2 does not belong to $\langle x \rangle y$ (as $y \notin \langle x \rangle$) and cannot equal x or x^3 (because the order of y is not 8). So if y has order 4, then $y^2 = x^2$. Hence we have four possibilities:

(i) If $yx = xy$ and $y^2 = e$, the group is abelian and $x \to (1,0)$, $y \to (0,1)$ leads to an isomorphism between G and $\mathbb{Z}_4 \times \mathbb{Z}_2$.

(ii) If $yx = x^3y$ and $y^2 = e$, then (with the usual notation) $x \to r$, $y \to s$ determines an isomorphism between G and D_4.

(iii) If $yx = xy$ and $y^2 = x^2$, the group is abelian, xy^{-1} has order 2, and $x \to (1,0)$, $xy^{-1} \to (0,1)$ leads to an isomorphism between G and $\mathbb{Z}_4 \times \mathbb{Z}_2$.

(iv) Finally, if $yx = x^3y$ and $y^2 = x^2$, then $x \to i$, $y \to j$ determines an isomorphism between G and Q.

What if every element of $G - \{e\}$ has order 2? In this case G is an abelian group. Choose x, y, z from $G - \{e\}$ and make sure xy is not equal to z. The subgroup $H = \{e, x, y, xy\}$ is isomorphic to $\mathbb{Z}_2 \times \mathbb{Z}_2$ and if $K = \langle z \rangle$ we easily check that $HK = G$ and $H \cap K = \{e\}$. Therefore, $G \cong H \times K \cong \mathbb{Z}_2 \times \mathbb{Z}_2 \times \mathbb{Z}_2$ by (10.2).   □

EXERCISES

13.1. Verify that the relation $\mathcal{R}$ used in the proof of Cauchy's theorem is indeed an equivalence relation.

13.2. If $p_1, p_2, \ldots, p_s$ are distinct primes, show that an abelian group of order $p_1p_2 \ldots p_s$ must be cyclic.

13.3. In Theorem 13.2 we showed that a group of order 6 must be cyclic or dihedral. Follow the proof until you have the six elements e, x, x^2, y, xy, x^2y. There are three possibilities for the order of xy. Show that one leads to $\mathbb{Z}_6$, a second to D_3, and the third to a contradiction.

13.4. Prove that a group of order 10 is either isomorphic to $\mathbb{Z}_{10}$ or isomorphic to D_5.

13.5. Let G be a group of order $4n + 2$. Use Cauchy's theorem, Cayley's theorem, and Exercise 6.6 to show that G contains a subgroup of order $2n + 1$.

13.6. Check that every proper subgroup of Q is cyclic.

13.7. Given quaternions $q = a + bi + cj + dk$, $q' = a' + b'i + c'j + d'k$ define

$$q + q' = (a + a') + (b + b')i + (c + c')j + (d + d')k,$$

$$q . q' = (aa' - bb' - cc' - dd') + (ab' + ba' + cd' - dc')i$$

$$+ (ac' - bd' + ca' + db')j + (ad' + bc' - cb' + da')k.$$

Prove that $\mathbb{H}$ forms an abelian group under addition, and that $\mathbb{H} - \{0\}$ is a group (though not an abelian group) under multiplication. Show that the correspondence

$$a + bi + cj + dk \leftrightarrow (a, b, c, d)$$

is an isomorphism from the additive group $\mathbb{H}$ to $\mathbb{R}^4$.

13.8. The conjugate of a quaternion $q = a + bi + cj + dk$ is defined to be $q^* = a - bi - cj - dk$, and the length of q is the square root of $q . q^*$; in other words $\sqrt{(a^2 + b^2 + c^2 + d^2)}$. Show that the quaternions of unit length form a subgroup of $\mathbb{H} - \{0\}$. We shall denote this group by S^3 because it corresponds to the unit sphere if we identify $\mathbb{H}$ with $\mathbb{R}^4$.

13.9. Prove that the correspondence

$$a + bi + cj + dk \leftrightarrow \begin{bmatrix} a + ib & c + id \\ -c + id & a - ib \end{bmatrix}$$

defines an isomorphism between S^3 and SU_2.

13.10. Write out the elements of SU_2 which correspond to the subgroup Q of S^3. Find a subgroup of S^3 which is isomorphic to C.

13.11. An element of $\mathbb{H}$ of the form $bi + cj + dk$ is called a pure quaternion. Show that $q . (bi + cj + dk) . q^{-1}$ is a pure quaternion for every $q \in \mathbb{H}$.

13.12. Given $\mathbf{x} = (x_1, x_2, x_3)$ in $\mathbb{R}^3$, let $q(\mathbf{x})$ denote the quaternion $x_1 i + x_2 j + x_3 k$. If $\mathbf{x}, \mathbf{y} \in \mathbb{R}^3$ prove that

$$q(\mathbf{x} \times \mathbf{y}) = \mathbf{x} . \mathbf{y} + q(\mathbf{x}) . q(\mathbf{y}).$$

CHAPTER 14

Conjugacy

The relation of conjugacy was introduced in Chapter 12 and shown to be an equivalence relation. We recall the definition. Given elements x, y of a group G we say that x *is conjugate to* y if $gxg^{-1} = y$ for some $g \in G$. The equivalence classes are called *conjugacy classes*, and we begin by working out these classes for some specific groups.

For a fixed element $g \in G$ the function from G to G given by $x \to gxg^{-1}$ is an isomorphism called **conjugation by g**. (It is a bijection because it is invertible, its inverse being conjugation by g^{-1}, and it preserves the algebraic structure of G because

$$g(xy)g^{-1} = (gxg^{-1})(gyg^{-1})$$

for any two elements $x, y \in G$.) Since an isomorphism preserves the order of an element we see that elements in the same conjugacy class must have the *same order*.

EXAMPLE (i). If G is abelian and if x is an element of G, then $gxg^{-1} = x$ for *all* $g \in G$. So x is only conjugate to itself and the conjugacy classes are the singletons $\{x\}$ where $x \in G$.

EXAMPLE (ii). Take G to be the dihedral group D_6 and adopt the notation of Chapter 4. The elements of D_6 are

$$e, r, r^2, r^3, r^4, r^5$$

$$s, rs, r^2 s, r^3 s, r^4 s, r^5 s$$

and multiplication is completely determined once we know $r^6 = e$, $s^2 = e$, $sr = r^5 s$. To find the conjugacy class of a power of r, say r^a where $1 \leqslant a \leqslant 5$, we must calculate $gr^a g^{-1}$ for every g in D_6. If g is the identity or a power of r, we get r^a back again. Taking $g = s$ (and remembering that $s^{-1} = s$) we have

$$sr^a s = r^{6-a} s^2 = r^{6-a}.$$

Finally, if $g = r^b s$ where $1 \leqslant b \leqslant 5$, then

$$(r^b s)r^a(r^b s)^{-1} = r^b(sr^a s)r^{6-b}$$
$$= r^b(r^{6-a})r^{6-b}$$
$$= r^{6-a}.$$

Therefore, the conjugacy class of r^a is $\{r^a, r^{6-a}\}$. For the remaining elements note that

$$r^b s r^{-b} = r^b r^b s = r^{2b} s$$

and

$$r^b(rs)r^{-b} = r^{b+1} r^b s = r^{2b+1} s.$$

Conjugation by $r^b s$ also sends s to $r^{2b} s$ and sends rs to $r^{2b-1} s$. Therefore, the elements $s, r^2 s, r^4 s$ form a conjugacy class, as do $rs, r^3 s, r^5 s$. In summary, the conjugacy classes of D_6 are

$$\{e\}, \{r, r^5\}, \{r^2, r^4\}, \{r^3\},$$
$$\{s, r^2 s, r^4 s\}, \{rs, r^3 s, r^5 s\}.$$

At this point we recommend Exercises 14.1 and 14.2.

EXAMPLE (iii). Two elements of S_n are said to have the same *cycle structure* if when they are decomposed as products of disjoint cyclic permutations they both have the same number of 2-cycles, the same number of 3-cycles, and so on. If $\theta, \varphi \in S_n$ have the same cycle structure, write out the cycle decomposition of φ underneath that of θ, taking the constituent cycles in order of decreasing length. In both cases include the integers left fixed by the permutation as cycles of length 1. Let g be the element of S_n which sends each integer mentioned in θ to the integer vertically below it in φ. Then $g\theta g^{-1} = \varphi$ because moving an integer up from φ to θ, pushing it along one position in θ, then dropping it back down to φ is the same as moving along one position in φ. *Therefore, permutations which have the same cycle structure are conjugate in S_n.*

Here is a specific calculation. The permutations $\theta = (67)(2539)(14)$, $\varphi = (12)(38)(5467)$ are both elements of S_9 and have the same cycle structure consisting of two transpositions plus a single 4-cycle. Our procedure gives

$$(2539)(67)(14)(8)$$
$$(5467)(12)(38)(9) \qquad \downarrow g$$

and we read off $g = (136)(254897)$. Thus,

$$g\theta g^{-1}(1) = g\theta(6)$$

$$= g(7)$$

$$= 2 = \varphi(1), \quad \text{etc.}$$

The element g is not unique. Writing θ as $(2539)(14)(67)(8)$ and keeping φ the same gives $g = (254)(36)(789)$.

Conversely, conjugate permutations have the same cycle structure. To see why, let $\theta = \theta_1 \theta_2 \ldots \theta_t$ be an element of S_n written as a product of disjoint cyclic permutations. For any $g \in S_n$ we have

$$g\theta g^{-1} = g(\theta_1 \theta_2 \ldots \theta_t)g^{-1}$$

$$= (g\theta_1 g^{-1})(g\theta_2 g^{-1})\ldots(g\theta_t g^{-1}).$$

Assume θ_i has length k, say $\theta_i = (a_1 a_2 \ldots a_k)$, then

$$g\theta_i g^{-1}(g(a_1)) = g\theta_i(a_1) = g(a_2),$$

$$g\theta_i g^{-1}(g(a_2)) = g\theta_i(a_2) = g(a_3),$$

$$\vdots$$

$$g\theta_i g^{-1}(g(a_k)) = g\theta_i(a_k) = g(a_1).$$

Also, if m is not one of $g(a_1), \ldots, g(a_k)$ then θ_i fixes $g^{-1}(m)$ and

$$g\theta_i g^{-1}(m) = gg^{-1}(m) = m.$$

Therefore, $g\theta_i g^{-1} = (g(a_1)g(a_2)\ldots g(a_k))$, a cyclic permutation of the same length as θ_i. Since $g\theta_1 g^{-1}, g\theta_2 g^{-1}, \ldots, g\theta_t g^{-1}$ are clearly disjoint, we conclude that $g\theta g^{-1}$ has the same cycle structure as θ.

EXAMPLE (iv). From the previous example we know that the conjugacy classes of S_4 are

$$\{\varepsilon\}$$

$$\{(12), (13), (14), (23), (24), (34)\},$$

$$\{(123), (132), (142), (124), (134), (143), (243), (234)\},$$

$$\{(1234), (1432), (1243), (1342), (1324), (1423)\},$$

$$\{(12)(34), (13)(24), (14)(23)\}.$$

How about those of A_4? We must be careful, if $\theta, \varphi \in A_4$ have the same cycle structure, there is certainly an element $g \in S_4$ such that $g\theta g^{-1} = \varphi$, but it may not be possible to produce an *even* permutation g with this property. For example if $g(123)g^{-1} = (132)$, then $(g(1)g(2)g(3)) = (132)$ and g must be one of the transpositions $(23), (13), (12)$. So g cannot lie in A_4. We quickly check that the conjugacy classes of A_4 are

$$\{\varepsilon\}$$

$$\{(123),(142),(134),(243)\},$$

$$\{(132),(124),(143),(234)\},$$

$$\{(12)(34),(13)(24),(14)(23)\}.$$

These classes have a simple *geometrical* interpretation. Identify A_4 with the rotational symmetry group of a regular tetrahedron in the usual way. Given an axis of symmetry through one of the vertices, we can rotate by $2\pi/3$ so that, when viewed from the vertex in question, the opposite face appears to move clockwise. The four rotations of this type are conjugate, as are the other four where the face moves anticlockwise. These classes correspond to the two distinct conjugacy classes of four 3-cycles. The identity rotation forms a conjugacy class on its own, and the remaining class consists of the three rotations through π about axes determined by the midpoints of pairs of opposite edges.

EXAMPLE (v). Take G to be O_2 and let

$$A_\theta = \begin{bmatrix} \cos\theta & -\sin\theta \\ \sin\theta & \cos\theta \end{bmatrix}, \qquad B_\varphi = \begin{bmatrix} \cos\varphi & \sin\varphi \\ \sin\varphi & -\cos\varphi \end{bmatrix}.$$

Remember that A_θ represents anticlockwise rotation through θ, and B_φ reflection in a line inclined at an angle of $\varphi/2$ to the positive x-axis. Conjugate matrices have the same determinant, so each conjugacy class will consist entirely of rotations or entirely of reflections. As

$$A_\theta B_\varphi A_\theta^{-1} = A_\theta B_\varphi A_{-\theta} = B_{\varphi+2\theta}$$

we see that any two of the B's are conjugate. Also

$$A_\varphi A_\theta A_\varphi^{-1} = A_\theta$$

and

$$B_\varphi A_\theta B_\varphi^{-1} = B_\varphi A_\theta B_\varphi = A_{-\theta}$$

which shows that the rotation matrices divide up into conjugacy classes of the form $\{A_\theta, A_{-\theta}\}$. Thus, the distinct conjugacy classes of O_2 are

$$\{I\},$$

$$\{A_\theta, A_{-\theta}\}, \qquad 0 < \theta < \pi,$$

$$\{A_\pi\},$$

$$\{B_\varphi | 0 \leqslant \varphi < 2\pi\}.$$

In the next chapter, we shall study subgroups which are made up of complete conjugacy classes. One such is the so-called centre of a group. The **centre** of a group G consists of all those elements which commute with every element of G. It is usually denoted by $Z(G)$ so that $Z(G) = \{x \in G | xg = gx, \forall g \in G\}$.

(14.1) Theorem. *The centre is a subgroup of G and is made up of the conjugacy classes which contain just one element.*

Proof. If $x, y \in Z(G)$ and $g \in G$, then

$$gxy^{-1} = xgy^{-1} \quad \text{(because } x \in Z(G))$$
$$= x(yg^{-1})^{-1}$$
$$= x(g^{-1}y)^{-1} \quad \text{(because } y \in Z(G))$$
$$= xy^{-1}g.$$

Therefore, xy^{-1} belongs to $Z(G)$. Certainly $e \in Z(G)$, so the centre is a subgroup by (5.1). Since $xg = gx$ if and only if $gxg^{-1} = x$, we see that x lies in $Z(G)$ precisely when the conjugacy class of x is the singleton $\{x\}$. $\qquad\square$

EXAMPLE (vi). The centre of any abelian group is the whole group.

EXAMPLE (vii). For $n \geqslant 3$ the centre of S_n is the trivial subgroup $\{\varepsilon\}$. This follows from Example (iii).

EXAMPLE (viii). The centre of D_6 is $\{e, r^3\}$ as we see from Example (ii). We ask the reader to work out the centre of D_n in the general case, distinguishing carefully between the cases n even and n odd (Exercise 14.10).

EXAMPLE (ix). The centre of GL_n consists of all (non-zero) scalar multiples of the identity matrix (Exercise 14.11).

EXERCISES

14.1. Work out the conjugacy classes of D_5.

14.2. Explain the structure of the conjugacy classes of D_n, distinguishing carefully between the cases where n is even and where n is odd.

14.3. Let $\varphi: G \to G'$ be an isomorphism. Prove that φ sends each conjugacy class of G to a conjugacy class of G'.

14.4. Calculate the number of different conjugacy classes in S_6 and write down a representative permutation for each class. Find an element $g \in S_6$ such that

$$g(123)(456)g^{-1} = (531)(264).$$

Show that $(123)(456)$ and $(531)(264)$ are conjugate in A_6, but $(12345)(678)$ and $(43786)(215)$ are *not* conjugate in A_8.

14.5. Prove that the 3-cycles in A_5 form a single conjugacy class. Find two 5-cycles in A_5 which are not conjugate in A_5.

14.6. How many elements of S_8 have the same cycle structure as $(12)(345)(678)$?

14.7. Work out the conjugacy classes and the centre of the quaternion group Q. What is the centre of S^3?

14.8. Use Exercise 6.10 to show that the centre of S_n is the trivial subgroup $\{\varepsilon\}$ when $n \geqslant 3$.

14.9. The group A_3 is abelian, therefore $Z(A_3) = A_3$. Prove that $Z(A_n) = \{\varepsilon\}$ when n is greater than 3.

14.10. Read off the centre of D_n from your calculations in Exercise 14.2. You should find that $Z(D_n)$ is $\{e\}$ when n is odd and $\{e, r^{n/2}\}$ when n is even.

14.11. Consider the matrices which can be obtained from the $n \times n$ identity matrix either by altering one of the diagonal entries to -1, or by interchanging two rows. Show that these matrices are all invertible and use them to compute the centre of $GL_n(\mathbb{R})$.

14.12. Find the centres of O_n and SO_n. Prove that the centre of U_n consists of all matrices of the form $e^{i\theta} I_n$ where $\theta \in \mathbb{R}$ and I_n denotes the $n \times n$ identity matrix.

CHAPTER 15

Quotient Groups

We shall devote this chapter to subgroups which are made up of *complete* conjugacy classes.

A subgroup H of a group G is called a **normal subgroup** *of G if H is a union of conjugacy classes of G.*

Normal subgroups are important because their left cosets form a group in a natural way. If X and Y are subsets of G, we can multiply them to form the set XY of all products xy where $x \in X$ and $y \in Y$.

(15.1) Theorem. *If H is a normal subgroup of G, the set of all left cosets of H in G forms a group under this multiplication.*

Proof. The product of two left cosets is again a left coset because

$$(xH)(yH) = xyH \qquad\qquad (*)$$

for any two elements $x, y \in G$. Accepting this for the moment, associativity follows from associativity in G, the coset $eH = H$ acts as an identity, and $x^{-1}H$ is the inverse of xH for each $x \in G$. So we do indeed have a group.

Just why does ($*$) hold and what does it have to do with the hypothesis that H be a *normal* subgroup of G? Each element of $(xH)(yH)$ has the form $xhyh'$ for some $h, h' \in H$. Rewrite this as

$$xy(y^{-1}hy)h'$$

and notice that $y^{-1}hy$ is a *conjugate* of h. By assumption the subgroup H is a normal subgroup of G and therefore contains the whole conjugacy class of h.

Hence, $y^{-1}hy = h''$ for some $h'' \in H$, giving

$$xhyh' = xy(y^{-1}hy)h' = xy(h''h').$$

Now we can see that $xhyh'$ belongs to xyH. So far we have $(xH)(yH) \subseteq xyH$. The reverse inclusion is easier to check and works for any subgroup H. Each element of xyH has the form xyh for some $h \in H$. Rewriting this as $(xe)(yh)$ shows that it belongs to $(xH)(yH)$ and we deduce $xyH \subseteq (xH)(yH)$. This completes the argument.   □

If H is a normal subgroup of G we write $H \lhd G$. The group of left cosets of H in G introduced above is called the **quotient group** (or factor group) of G by H and denoted G/H. We recall that the left cosets of H in G form a partition of G. Each of these cosets represents a *single element* in G/H and it is in this sense that we have "divided G by H."

EXAMPLE (i). Let H be the subgroup of D_6 generated by the element r^3; then H is a normal subgroup of D_6 as it is made up of the conjugacy classes $\{e\}$ and $\{r^3\}$. There are six distinct left cosets, namely

$$eH = \{e, r^3\}, \qquad rH = \{r, r^4\}, \qquad r^2H = \{r^2, r^5\},$$
$$sH = \{s, sr^3\} = \{s, r^3s\}, \qquad rsH = \{rs, r^4s\},$$

and

$$r^2sH = \{r^2s, r^5s\}.$$

These are the *elements* of D_6/H. Our definition of multiplication gives

$$(rH)(sH) = \{xy | x \in rH, y \in sH\}$$
$$= \{rs, rr^3s, r^4s, r^4r^3s\}$$
$$= \{rs, r^4s\}$$
$$= rsH$$

exactly as predicted by (∗) in (15.1). Writing our cosets as

$$eH, rH, (rH)^2, sH, (rH)(sH), (rH)^2(sH)$$

and checking

$$(rH)^3 = r^3H = eH,$$
$$(sH)^2 = s^2H = eH,$$
$$(sH)(rH) = srH = r^2sH = (rH)^2(sH)$$

we conclude that the quotient group D_6/H is isomorphic to D_3.

EXAMPLE (ii). The conjugacy classes $\{\varepsilon\}$ and $\{(12)(34), (13)(24), (14)(23)\}$ make up a normal subgroup J of A_4. There are three left cosets εJ, $(123)J$, $(132)J$, and the quotient group A_4/J is isomorphic to $\mathbb{Z}_3$.

EXAMPLE (iii). Every subgroup of an abelian group is a normal subgroup because in this case the conjugacy classes are just the elements of the group. Let n be a positive integer and consider the subgroup $n\mathbb{Z}$ (all multiples of n) of $\mathbb{Z}$. There are n distinct cosets

$$0 + n\mathbb{Z}, 1 + n\mathbb{Z}, \ldots, (n-1) + n\mathbb{Z}$$

which when combined via

$$(x + n\mathbb{Z}) + (y + n\mathbb{Z}) = (x + y) + n\mathbb{Z}$$

make up the quotient group $\mathbb{Z}/n\mathbb{Z}$. The element $1 + n\mathbb{Z}$ generates the whole group; therefore, $\mathbb{Z}/n\mathbb{Z}$ is isomorphic to $\mathbb{Z}_n$.

If we wish to check that the subgroup H of G is normal, without first working out all the conjugacy classes of G, we can do so by showing that

$$ghg^{-1} \in H \quad \text{for all } h \in H, g \in G.$$

For example, SO_n is a normal subgroup of O_n because every conjugate of a matrix of determinant $+1$ also has determinant $+1$. We can sometimes be more efficient using the following result.

(15.2) Theorem. *Let H be a subgroup of G and let X be a set of generators for G which, if G is infinite, contains the inverse of each of its elements. Then H is a normal subgroup of G, provided $xhx^{-1} \in H$ for all $h \in H$, $x \in X$.*

Proof. If $h \in H$, we must show that every conjugate ghg^{-1} also belongs to H. Express g as a product $x_1 x_2 \ldots x_t$ in which each x_i is an element of X. Then

$$ghg^{-1} = (x_1 x_2 \ldots x_t)h(x_1 x_2 \ldots x_t)^{-1}$$
$$= x_1 x_2 \ldots x_t h x_t^{-1} \ldots x_2^{-1} x_1^{-1}$$

and conjugation by g amounts to repeated conjugation by elements of X. By assumption, each time we conjugate by an element of X we produce another member of H. Therefore, ghg^{-1} does indeed belong to H. ☐

EXAMPLE (iv). The subgroup H of D_n generated by r^2 is a normal subgroup. Take $X = \{r, s\}$ and check that conjugating a power of r^2 by either r or s gives back a power of r^2. If n is *odd*, then $H = \langle r^2 \rangle = \langle r \rangle$, so D_n/H has two elements H, sH, and is isomorphic to $\mathbb{Z}_2$. If n is *even* there are four distinct cosets H, rH, sH, rsH. We note that

$$(rH)^2 = r^2 H = H \qquad (\text{because } r^2 \in H),$$
$$(sH)^2 = s^2 H = H,$$

and

$$(rsH)^2 = (rs)^2 H = H.$$

Therefore, we have a group of order 4 in which the square of every element is the identity. Hence, D_n/H is isomorphic to $\mathbb{Z}_2 \times \mathbb{Z}_2$.

We have chosen to work with left cosets, but the next result shows that H is a normal subgroup of G precisely when its left and right cosets coincide.

(15.3) Theorem. *The subgroup H of G is normal if and only if $xH = Hx$ for all $x \in G$.*

Proof. Suppose H is normal. Given $x \in G, h \in H$, we know that the conjugates xhx^{-1} and $x^{-1}hx$ must belong to H. Therefore,

$$xh = (xhx^{-1})x \in Hx$$

and we have $xH \subseteq Hx$. Similarly, $hx = x(x^{-1}hx) \in xH$ provides the reverse inclusion $Hx \subseteq xH$. Now assume $xH = Hx$ for all $x \in G$. If $h \in H, x \in G$, the conjugate xhx^{-1} belongs to

$$(xH)x^{-1} = (Hx)x^{-1} = H,$$

and therefore H must be normal in G.   □

This means that we can equally well think of the elements of G/H as the right cosets of H in G, multiplying them via $(Hx)(Hy) = Hxy$. The **index** of a subgroup H of G is the number of distinct left (or right) cosets of H in G. Even if H is not a normal subgroup it does not matter which type of coset we use, since the correspondence $xH \rightarrow Hx^{-1}$ is a bijection between the set of left cosets and the set of right cosets.

(15.4) Theorem. *If the index of H in G is equal to 2, then H is a normal subgroup of G and the quotient group G/H is isomorphic to $\mathbb{Z}_2$.*

Proof. We show that $xH = Hx$ for every $x \in G$. This is clear when $x \in H$. If $x \in G - H$, the left cosets H, xH form a partition of G. But H and Hx also partition G. Therefore, xH must equal Hx as required. As there are only two distinct cosets, the quotient group G/H has order 2 and is isomorphic to $\mathbb{Z}_2$.
□

EXAMPLE (v). A_n is a normal subgroup of S_n.

EXAMPLE (vi). The subgroup generated by r is a normal subgroup of D_n.

EXAMPLE (vii). For a second time we see that SO_n is a normal subgroup of O_n. The index of SO_n in O_n is equal to two because the cosets ISO_n, USO_n fill out O_n, where I is the identity matrix and U is obtained from I by changing the final 1 on the diagonal to -1.

(15.5) Theorem. *If p is an odd prime, any group of order $2p$ is either cyclic or dihedral.*

Proof. Use Cauchy's theorem to select an element x of order p and an element y of order 2. The right cosets $\langle x \rangle$, $\langle x \rangle y$ provide $2p$ elements $e, x, \ldots, x^{p-1}$,

y, xy, $\ldots x^{p-1}y$, which fill out the group, and $\langle x \rangle$ is a normal subgroup because its index is equal to 2. By Lagrange's theorem the order of xy is $2p$, p, or 2. If xy has order $2p$, then our group is cyclic. If it has order 2, then $xyxy = e$, giving $yx = x^{-1}y$, and we have a group which is isomorphic to the dihedral group D_p. We claim that xy cannot have order p. For suppose $(xy)^p = e$, then

$$\langle x \rangle = \langle x \rangle (xy)^p = (\langle x \rangle xy)^p = (\langle x \rangle y)^p = \langle x \rangle y^p = \langle x \rangle y,$$

which leads to the contradiction $y \in \langle x \rangle$.   □

We now change tack and describe a process which allows us to "abelianise" an arbitrary group G. An element of G of the form $xyx^{-1}y^{-1}$ is called a *commutator* and the subgroup generated by all commutators is the *commutator subgroup* $[G, G]$ of G. Since x and y commute precisely when the commutator $xyx^{-1}y^{-1}$ is the identity, the size of $[G, G]$ can be thought of as a measure of how far G is from being abelian. The commutator subgroup of any abelian group is just the trivial subgroup $\{e\}$.

EXAMPLE (viii). Each commutator in S_n is obviously an even permutation and therefore $[S_n, S_n]$ is contained in A_n. Every 3-cycle is a commutator because

$$(abc) = (ab)(ac)(ab)(ac),$$

and the 3-cycles generate A_n when $n \geqslant 3$. We conclude that the commutator subgroup of S_n is all of A_n.

(15.6) Theorem. *The commutator subgroup is a normal subgroup, the quotient group $G/[G, G]$ is abelian, and if H is a normal subgroup of G for which G/H is abelian, then $[G, G]$ is contained in H.*

Proof. Any conjugate of a commutator is again a commutator because

$$g(xyx^{-1}y^{-1})g^{-1} = (gxg^{-1})(gyg^{-1})(gxg^{-1})^{-1}(gyg^{-1})^{-1}.$$

A general element of $[G, G]$ may not be a commutator, but is a product of commutators, say $c_1 c_2 \ldots c_k$. Conjugating by an element of G gives

$$g(c_1 c_2 \ldots c_k)g^{-1} = (gc_1 g^{-1})(gc_2 g^{-1}) \ldots (gc_k g^{-1})$$

which again lies in $[G, G]$. Therefore, the commutator subgroup is a normal subgroup of G.

If x, y are elements of G, then $xyx^{-1}y^{-1} \in [G, G]$. Hence, $[G, G]xyx^{-1}y^{-1} = [G, G]$, which in turn gives $[G, G]xy = [G, G]yx$ and shows that $G/[G, G]$ is abelian.

Finally, if G/H is abelian and if $x, y \in G$, then $Hxy = Hyx$. Therefore, $Hxyx^{-1}y^{-1} = H$, which tells us that the commutator $xyx^{-1}y^{-1}$ must lie in H. So H contains $[G, G]$.   □

The commutator subgroup is the *smallest* normal subgroup of G for which the corresponding quotient group is abelian. In forming $G/[G, G]$ we say that we **abelianise** G.

EXAMPLE (ix). With our usual notation $\langle r^2 \rangle$ is a normal subgroup of D_n. If n is odd, then $\langle r^2 \rangle = \langle r \rangle$ has index 2 in D_n, whereas for n even the index of $\langle r^2 \rangle$ in D_n is 4. In both cases the quotient group $D_n/\langle r^2 \rangle$ must be abelian and therefore the commutator subgroup of D_n is contained in $\langle r^2 \rangle$. But

$$rsr^{-1}s^{-1} = rrss = r^2$$

showing that r^2 is a commutator. Hence, $[D_n, D_n] = \langle r^2 \rangle$. If n is odd, we have $D_n/[D_n, D_n] \cong \mathbb{Z}_2$, and for n even, $D_n/[D_n, D_n] \cong \mathbb{Z}_2 \times \mathbb{Z}_2$.

EXAMPLE (x). The subgroup $\{\pm 1\}$ of the quaternion group Q is a normal subgroup and $Q/\{\pm 1\}$ is isomorphic to $\mathbb{Z}_2 \times \mathbb{Z}_2$. Therefore, the commutator subgroup of Q is contained in $\{\pm 1\}$, and it must equal $\{\pm 1\}$ because Q is not an abelian group.

EXERCISES

15.1. If H and J are both subgroups of a group G, prove that HJ is a subgroup of G if and only if $HJ = JH$.

15.2. Find all normal subgroups of D_4 and D_5. Generalise and deal with D_n for arbitrary n.

15.3. Show that every subgroup of the quaternion group Q is a normal subgroup of Q.

15.4. Is O_n a normal subgroup of $GL_n(\mathbb{R})$?

15.5. Let H be a normal subgroup of a group G, and let J be a normal subgroup of H. Then of course J is a subgroup of G. Supply an example to show that J need not be normal in G.

15.6. If H, J are normal subgroups of a group, and if they have only the identity element in common, show that $xy = yx$ for all $x \in H$, $y \in J$.

15.7. Let K be a normal subgroup of $G \times H$, which has only the identity in common with each of $G \times \{e\}$ and $\{e\} \times H$. Show that K is abelian.

15.8. Find the commutator subgroup of A_4. If n is at least 5, show that the commutator subgroup of A_n is all of A_n.

15.9. Let $\varphi: G \to G'$ be an isomorphism. Prove that φ sends the commutator subgroup of G to the commutator subgroup of G'.

15.10. Improve Theorem 15.2 as follows: Let H be a subgroup of a group G, let X be a set of generators for G which contains the inverse of each of its elements, and let Y be a set of generators for H. Show that H is a normal subgroup of G if $xyx^{-1} \in H$ for all $x \in X$, $y \in Y$.

15.11. Let H be a subgroup of finite index of an infinite group G. Prove that G has a normal subgroup of finite index which is contained in H.

15.12. A *simple group* is one whose only normal subgroups are $\{e\}$ and the whole group. Find a proper normal subgroup of A_4. Now consider the alternating group A_5. Work out the commutators

$$(12345)^{-1}(345)^{-1}(12345)(345),$$

$$(12)(34)(345)^{-1}(12)(34)(345).$$

Show that a non-trivial normal subgroup of A_5 must contain a 3-cycle. Use the first part of Exercise 14.5 to conclude that this subgroup must be *all of A_5*, and that A_5 is therefore a simple group. (It is not much harder to verify that A_n is simple when n is greater than 5.)

15.13. If H is a *cyclic* normal subgroup of a group G, prove that any subgroup of H is also a normal subgroup of G.

15.14. Show that every element of the quotient group $\mathbb{Q}/\mathbb{Z}$ has finite order, but that only the identity element of $\mathbb{R}/\mathbb{Q}$ has finite order.

15.15. If G contains a normal subgroup which is isomorphic to $\mathbb{Z}_2$, and if the corresponding quotient group is infinite cyclic, prove that G is isomorphic to $\mathbb{Z} \times \mathbb{Z}_2$.

15.16. Suppose that G contains an infinite cyclic normal subgroup for which the corresponding quotient group is cyclic of order 2. Show that G must be isomorphic to one of $\mathbb{Z}$, $\mathbb{Z} \times \mathbb{Z}_2$, D_∞.

CHAPTER 16

Homomorphisms

Let G, G' be groups. A function $\varphi: G \to G'$ is a **homomorphism** if it takes the multiplication of G to that of G'; in other words if

$$\varphi(xy) = \varphi(x)\varphi(y) \quad \text{for all } x, y \in G.$$

The **kernel** K of φ is then defined to be the set of those elements of G which φ maps to the identity of G'; in symbols $K = \{x \in G | \varphi(x) = e\}$. If φ is also a bijection, then it is an isomorphism, and in this case its kernel is just the identity element of G. Various properties of isomorphisms were checked in Chapter 7. Those arguments which do not use the fact that an isomorphism is a bijection are equally valid here. Therefore, a homomorphism sends the identity of G to that of G', sends inverses to inverses, and sends each subgroup of G to a subgroup of G'. In particular, φ maps the whole group G to a subgroup of G' which is called the **image** of φ.

Notice that if H is a normal subgroup of G the function $\varphi: G \to G/H$ defined by $\varphi(x) = xH$ is a homomorphism because

$$\varphi(xy) = xyH = (xH)(yH) = \varphi(x)\varphi(y)$$

for all $x, y \in G$. The image of this homomorphism is G/H and its kernel is precisely H.

(16.1) First Isomorphism Theorem. *The kernel K of a homomorphism $\varphi: G \to G'$ is a normal subgroup of G, and the correspondence $xK \to \varphi(x)$ is an isomorphism from the quotient group G/K to the image of φ.*

Proof. Suppose $x, y \in K$, then $\varphi(xy^{-1}) = \varphi(x)\varphi(y)^{-1} = e$, showing that $xy^{-1} \in K$. Certainly K is non-empty because $e \in K$, hence K is a subgroup of G by (5.1). If $x \in K$ and $g \in G$, then

$$\varphi(gxg^{-1}) = \varphi(g)\varphi(x)\varphi(g)^{-1} = \varphi(g)\varphi(g)^{-1} = e.$$

Therefore, gxg^{-1} belongs to K and the subgroup K is normal in G.

If two cosets xK, yK are equal, then $y^{-1}x \in K$. Applying φ gives $\varphi(y^{-1}x) = \varphi(y)^{-1}\varphi(x) = e$, and therefore $\varphi(x) = \varphi(y)$. This means we have a function $\psi: G/K \to G'$ defined by $\psi(xK) = \varphi(x)$. Reversing the above computation shows that if $\varphi(x) = \varphi(y)$, then $xK = yK$, so ψ is injective. It is a homomorphism because

$$\psi(xKyK) = \psi(xyK) = \varphi(xy) = \varphi(x)\varphi(y) = \psi(xK)\psi(yK)$$

for any two cosets xK, $yK \in G/K$. Finally, the image of ψ is the same as the image of φ. We have proved that ψ is an isomorphism from G/K to the image of φ. $\qquad\square$

Two special cases of (16.1) are particularly useful.

(16.2) Corollary. *If the image of φ is all of G', then G/K is isomorphic to G'.*

(16.3) Corollary. *Suppose the image of φ is all of G'. Then φ is an isomorphism if and only if K consists just of the identity element of G.*

EXAMPLES. The reader should check that each of the following functions is a surjective homomorphism.

(i) $\mathbb{Z} \to \mathbb{Z}_n$, $\quad x \to x(\text{mod } n)$.
   $K = n\mathbb{Z}$ (the set of all multiples of n) and $\mathbb{Z}/n\mathbb{Z}$ is isomorphic to $\mathbb{Z}_n$.

(ii) $\mathbb{R} \to C$, $\quad x \to e^{2\pi i x}$.
   $K = \mathbb{Z}$ and $\mathbb{R}/\mathbb{Z}$ is isomorphic to C.

(iii) $\mathbb{C} - \{0\} \to C$, $\quad z \to z/|z|$.
   $K = \mathbb{R}^{\text{pos}}$ and $\mathbb{C} - \{0\}/\mathbb{R}^{\text{pos}}$ is isomorphic to C.

(iv) $O_n \to \{\pm 1\}$, $\quad A \to \det A$.
   $K = SO_n$ and O_n/SO_n is isomorphic to $\mathbb{Z}_2$.

(v) $U_n \to C$, $\quad A \to \det A$.
   $K = SU_n$ and U_n/SU_n is isomorphic to C.

(vi) $C \to C$, $\quad z \to z^2$.
   $K = \{\pm 1\}$ and $C/\{\pm 1\}$ is isomorphic to C.

(vii) The group S_4 contains three elements of order 2, namely $(12)(34)$, $(13)(24), (14)(23)$. Together with the identity, these elements form a subgroup of S_4 which is isomorphic to Klein's group and which we denote by V. Conjugation by a permutation $\theta \in S_4$ must permute our *three* elements of order 2 among themselves, because conjugate elements always have the same order. By sending each θ to the corresponding permutation (of these elements of order 2) we can produce a function from S_4 to S_3 which is a homomorphism and surjective. Its kernel is precisely V and (16.2) shows that S_4/V is isomorphic to S_3.

(viii) An element of $\mathbb{H}$ of the form $bi + cj + dk$ is called a "pure quaternion." Identify the set of all pure quaternions with $\mathbb{R}^3$ via the correspondence $bi + cj + dk \to (b, c, d)$. If q is a non-zero quaternion, conjugation by q sends the pure quaternions to themselves and induces a *rotation* of $\mathbb{R}^3$. This construction provides a homomorphism from $\mathbb{H} - \{0\}$ to SO_3. Its image is all of SO_3, its kernel is $\mathbb{R} - \{0\}$, and (16.2) tells us that $\mathbb{H} - \{0\}/\mathbb{R} - \{0\}$ is isomorphic to SO_3.

(16.4) Second Isomorphism Theorem. *Suppose H, J are subgroups of G with J normal in G. Then HJ is a subgroup of G, $H \cap J$ is a normal subgroup of H, and the quotient groups HJ/J, $H/H \cap J$ are isomorphic.*

Proof. Let g, g_* be elements of HJ and write $g = xy$, $g_* = x_* y_*$ where $x, x_* \in H$, $y, y_*, \in J$. Then

$$gg_*^{-1} = xyy_*^{-1}x_*^{-1}$$

$$= (xx_*^{-1})(x_*yy_*^{-1}x_*^{-1}) \in HJ,$$

and HJ is a subgroup of G by (5.1). Where has the normality of J been used?

The function $\varphi: H \to HJ/J$ defined by $\varphi(x) = xeJ = xJ$ is a homomorphism. It is surjective because if $g = xy \in HJ$, then

$$\varphi(x) = xJ$$

$$= xyJ \quad \text{(because } y \in J)$$

$$= gJ.$$

The element x of H belongs to the kernel of φ precisely when $xJ = J$, in other words when $x \in J$. Therefore, the kernel of φ is $H \cap J$ and the result follows from (16.1). $\square$

(16.5) Third Isomorphism Theorem. *Let H, J be normal subgroups of G and suppose H is contained in J. Then J/H is a normal subgroup of G/H and the quotient group $(G/H)/(J/H)$ is isomorphic to G/J.*

Proof. The function $\varphi: G/H \to G/J$ defined by $\varphi(xH) = xJ$ is a homomorphism and is surjective. A coset xH belongs to the kernel of φ precisely when $xJ = J$; in other words, when $x \in J$. Therefore, the kernel of φ is J/H and the result follows from (16.1). $\square$

EXERCISES

16.1. Which of the following define homomorphisms from $\mathbb{C} - \{0\}$ to $\mathbb{C} - \{0\}$?
   (a) $z \to z^*$ (b) $z \to z^2$
   (c) $z \to iz$ (d) $z \to |z|$

16.2. Do any of the following determine homomorphisms from $GL_n(\mathbb{C})$ to $GL_n(\mathbb{C})$?
(a) $A \to A^t$ (b) $A \to (A^{-1})^t$
(c) $A \to A^2$ (d) $A \to A^*$

16.3. Show that $G \times \{e\}$ is a normal subgroup of $G \times H$ and that the quotient group $(G \times H)/(G \times \{e\})$ is isomorphic to H.

16.4. If A is a normal subgroup of G, and if B is a normal subgroup of H, prove that $A \times B$ is a normal subgroup of $G \times H$ and that the quotient group $(G \times H)/(A \times B)$ is isomorphic to $(G/A) \times (H/B)$.

16.5. Verify that the translations in D_∞ form a normal subgroup of D_∞ and that the corresponding quotient group is isomorphic to $\mathbb{Z}_2$.

16.6. Given numbers $a \in \mathbb{R} - \{0\}$, $b \in \mathbb{R}$ define a function $f(a, b)$ from $\mathbb{R}$ to $\mathbb{R}$ by $f(a, b)(x) = ax + b$. Show that the collection of all such functions forms a group G under composition of functions. If H consists of those elements of G for which $a = 1$, prove that H is a normal subgroup of G and that G/H is isomorphic to $\mathbb{R} - \{0\}$.

16.7. Each element

$$\begin{bmatrix} a & b \\ c & d \end{bmatrix}$$

of $GL_2(\mathbb{C})$ gives rise to a so called **Möbius transformation**

$$z \to \frac{az + b}{cz + d}$$

of the extended complex plane $\mathbb{C} \cup \{\infty\}$. Show that these transformations form a group, the **Möbius group**, under composition of functions, and that this group is isomorphic to the quotient of $GL_2(\mathbb{C})$ by its centre.

16.8. Show that a function $\varphi: G \to G'$ is a homomorphism if and only if $\{(g, \varphi(g)) | g \in G\}$ is a subgroup of $G \times G'$.

16.9. Prove that the quotient group $S^3/\{\pm I\}$ is isomorphic to SO_3. Check that S^3 and SO_3 are *not* isomorphic to one another.

16.10. Use the idea of Exercise 12.12 to argue that S_n is isomorphic to a quotient of the braid group B_n.

16.11. Let $\varphi: G \to G'$ be a surjective homomorphism and let K denote its kernel. If H' is a subgroup of G' define

$$\varphi^{-1}(H') = \{g \in G | \varphi(g) \in H'\}.$$

Verify that $\varphi^{-1}(H')$ is a subgroup of G which contains K, and that the correspondence $H' \leftrightarrow \varphi^{-1}(H')$ is a *bijection* between the collection of all subgroups of G' and the collection of all those subgroups of G which contain K.

16.12. Show that H is a maximal normal subgroup of G if and only if G/H is a simple group. (Maximal means that the only normal subgroups of G which contain H are G and H.)

CHAPTER 17

Actions, Orbits, and Stabilizers

A good definition should be precise, economical, and capture a simple intuitive idea. If in addition it is easy to work with, so much the better. We begin this chapter with a definition having all these qualities.

*An **action** of a group G on a set X is a homomorphism from G to S_X.*

Examine this in detail. Let $\varphi\colon G \to S_X$ be a homomorphism. For each group element g in G the function φ gives us a permutation $\varphi(g)$ of the points of X. If $g,h \in G$ the permutation $\varphi(gh)$ associated to the product gh is equal to the composition $\varphi(g)\varphi(h)$ because φ is a homomorphism. We imagine the elements of G permuting the points of X in a way that is *compatible* with the algebraic structure of G.

Most of the time we will simplify our notation by writing $g(x)$ for the image of the point $x \in X$ under the permutation corresponding to g, instead of the more cumbersome $\varphi(g)(x)$. We shall often say that the group element $g \in G$ *sends* the point $x \in X$ to the point $g(x) \in X$. The requirement that φ be a homomorphism becomes

$$gh(x) = g(h(x))$$

for any two elements $g,h \in G$ and any point $x \in X$.

EXAMPLE (i). The infinite cyclic group $\mathbb{Z}$ acts on the real line by translation. The integer n sends the real number x to $n + x$. If m and n are integers, then

$$(m + n) + x = m + (n + x),$$

so we do have an action.

EXAMPLE (ii). Even the first word of our definition is suggestive. We say "an" action because a given group may act on the same set in different ways. Here is a second action of $\mathbb{Z}$ on $\mathbb{R}$. Agree that $n \in \mathbb{Z}$ sends $x \in \mathbb{R}$ to $(-1)^n x$. In other words, the permutation associated to every even integer is the identity permutation of $\mathbb{R}$, and that associated to all the odd integers is $x \to -x$. Since $(-1)^{m+n}x = (-1)^m(-1)^n x$ for any two integers m, n and any real number x, this is an action.

Before increasing our stock of examples, we need a little more terminology. Given an action of G on X and a point $x \in X$, the set of all images $g(x)$, as g varies through G, is called the **orbit** of x and written $G(x)$. Notice that $G(x)$ is a subset of X. In our first example the orbit of a real number x consists of all translates $n + x$ where $n \in \mathbb{Z}$. In the second example the orbit of x consists of the two points x, $-x$ provided x is not zero, and the orbit of 0 is just $\{0\}$.

Let $\mathscr{R}$ be the subset of $X \times X$ consisting of those ordered pairs (x, y) such that $g(x) = y$ for some $g \in G$. Then $\mathscr{R}$ is an equivalence relation on X. (The permutation associated to the identity element of G is the identity permutation, consequently each x is related to itself because $e(x) = x$. If x is related to y, say $g(x) = y$, then

$$g^{-1}(y) = g^{-1}(g(x)) = g^{-1}g(x) = e(x) = x$$

and y is related to x. If x is related to y, and y to z, say $g(x) = y$, $g'(y) = z$, then

$$g'g(x) = g'(g(x)) = g'(y) = z$$

showing that x is related to z.) The equivalence classes of $\mathscr{R}$ are just the orbits of our group action. Therefore, the distinct orbits partition X.

If x is a point of X, the elements of G which leave x fixed form a subgroup of G called the **stabilizer** G_x of x. In Example (i) the stabilizer of each real number is the trivial subgroup $\{0\}$ of $\mathbb{Z}$. In Example (ii) the stabilizer of x is the subgroup $2\mathbb{Z}$ of $\mathbb{Z}$ if x is not zero, and all of $\mathbb{Z}$ when $x = 0$.

EXAMPLE (iii). Let X be the set of edges of a cube. We can produce an action of $\mathbb{Z}_4$ on X by rotating the cube about an *axis* which passes through the centres of two opposite faces. Formally, if r is the permutation of X induced by the rotation shown in Figure 17.1, then we define $\varphi : \mathbb{Z}_4 \to S_X$ by $\varphi(m) = r^m$. There are three distinct orbits; the top four edges, the bottom four edges, and the four vertical edges. The stabilizer of every edge is the trivial subgroup $\{0\}$ of $\mathbb{Z}_4$.

EXAMPLE (iv). Orthogonal transformations preserve length, so if X is the unit sphere in $\mathbb{R}^3$, we have a natural action of SO_3 on X. The matrix $A \in SO_3$ sends the unit vector $\mathbf{x}$ to the unit vector $\mathbf{x}A^t$. The orbit of any vector is the *whole sphere*. (Two unit vectors $\mathbf{x}$, $\mathbf{y}$ determine a plane and a suitable rotation of $\mathbb{R}^3$ about the axis through $\mathbf{0}$ perpendicular to this plane will bring $\mathbf{x}$ to the position of $\mathbf{y}$. The matrix of this rotation is an element of SO_3 which sends $\mathbf{x}$ to $\mathbf{y}$.) Let

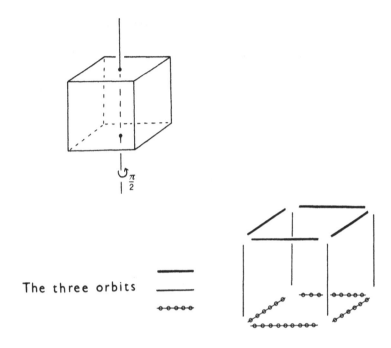

The three orbits

Figure 17.1

e_1 be the first member of the standard basis for $\mathbb{R}^3$. If $A \in SO_3$ fixes e_1, then A has the form

$$\begin{bmatrix} 1 & 0 & 0 \\ 0 & & \\ 0 & & B \end{bmatrix},$$

where $B \in SO_2$. Thus, the stabilizer of e_1 is this "standard copy" of SO_2 inside SO_3. Points in the same orbit always have conjugate stabilizers (see (17.1)); therefore, the stabilizer of every unit vector is isomorphic to SO_2. This is an example of a transitive action. An action is *transitive* if there is only one orbit.

EXAMPLE (v). Take any group G and let X be the underlying set of G. Let G act on X by conjugation. Here $g \in G$ sends x to gxg^{-1}. This is an action because

$$(gh)x(gh)^{-1} = g(hxh^{-1})g^{-1}$$

for all $g, h \in G$ and all $x \in X$. The orbits are the *conjugacy classes* of G. The stabilizer of the point x is

$$\{g \in G | gxg^{-1} = x\} = \{g \in G | gx = xg\},$$

in other words, the subgroup of G consisting of all those elements which *commute* with x.

Suppose we are given an action of a group G on a set X.

(17.1) Theorem. *Points in the same orbit have conjugate stabilizers.*

Proof. Suppose x and y belong to the same orbit, say $g(x) = y$. We show that

$$gG_x g^{-1} = G_y.$$

If $h \in G_x$, then

$$ghg^{-1}(y) = ghg^{-1}(g(x))$$

$$= gh(x)$$

$$= g(x)$$

$$= y.$$

Therefore, $gG_x g^{-1}$ is contained in G_y. Reversing the roles of x and y the same argument provides $g^{-1}G_y g \subseteq G_x$; in other words, $G_y \subseteq gG_x g^{-1}$. This completes the proof. $\qquad\square$

(17.2) Orbit-Stabilizer Theorem. *For each $x \in X$, the correspondence $g(x) \rightarrow gG_x$ is a bijection between $G(x)$ and the set of left cosets of G_x in G.*

Proof. The correspondence is clearly surjective. It is injective because if $gG_x = g'G_x$, then $g = g'h$ for some element h of G_x, and therefore $g(x) = g'h(x) = g'(h(x)) = g'(x)$. $\qquad\square$

Put another way, the Orbit-Stabilizer theorem says that the cardinality of the orbit of x is equal to the index of the stabilizer of x in G.

(17.3) Corollary. *If G is finite, the size of each orbit is a divisor of the order of G.*

Proof. By the Orbit-Stabilizer theorem the size of $G(x)$ is $|G|/|G_x|$, therefore

$$|G(x)| \cdot |G_x| = |G|.$$

$\qquad\square$

EXAMPLE (v, continued). If G is finite, (17.3) tells us that the number of elements in each conjugacy class is a divisor of the order of G.

EXAMPLE (vi). We can recast the proof of Cauchy's theorem as follows. Let G be a finite group and p be a prime divisor of the order of G. As before, X is the set of all ordered strings $(x_1, x_2, \ldots, x_p)$ of elements of G for which $x_1 x_2 \ldots x_p = e$. We recall that the size of X is divisible by p. There is a natural action of $\mathbb{Z}_p$ on X. The element $m \in \mathbb{Z}_p$ sends the string $(x_1, x_2, \ldots, x_p)$ to $(x_{m+1}, \ldots, x_p, x_1, \ldots, x_m)$. By (17.3) each orbit contains either 1 or p strings. If all orbits other than that of $(e, e, \ldots, e)$ contain p strings, then the size of

X cannot be divisible by p. We therefore have a string $(x_1, x_2, \ldots, x_p)$ other than $(e, e, \ldots, e)$ which is *left fixed* by every element of $\mathbb{Z}_p$. In other words, $x_1 = x_2 = \cdots = x_p$ as required.

(17.4) Theorem. *If p is prime and if the order of G is a power of p, then G has a non-trivial centre.*

Proof. Consider the partition of G into its conjugacy classes. We know from (17.3) that the size of each class is either 1 or a power of p, and by definition the centre is made up of the conjugacy classes which contain just one element. If the centre was trivial, the order of G would be congruent to 1 modulo p, contradicting our assumption that the order of G is a power of p. □

(17.5) Theorem. *If p is prime a group of order p^2 is either cyclic or isomorphic to $\mathbb{Z}_p \times \mathbb{Z}_p$.*

Proof. Suppose the order of G is p^2. If G contains an element of order p^2, then it is cyclic. Otherwise, all the elements except e have order p. The centre of G is non-trivial by (17.4), so choose x (not e) from $Z(G)$ and choose y outside $\langle x \rangle$. The p^2 elements $x^i y^j$, $1 \leqslant i, j \leqslant p$, are distinct because $\langle x \rangle$ and $\langle y \rangle$ only have the identity element in common. Therefore, $\langle x \rangle \langle y \rangle = G$. Every element of $\langle x \rangle$ commutes with every element of $\langle y \rangle$ since $x \in Z(G)$. By (10.2), G is isomorphic to $\langle x \rangle \times \langle y \rangle$ and hence to $\mathbb{Z}_p \times \mathbb{Z}_p$. □

EXERCISES

17.1. Let G be the subgroup of S_8 generated by $(123)(45)$ and (78). Then G acts as a group of permutations of the set $X = \{1, 2, \ldots, 8\}$. Calculate the orbit and the stabilizer of every integer in X.

17.2. The infinite dihedral group D_∞ acts on the real line in a natural way (see Chapter 5). Work out the orbit and the stabilizer of each of the points $1, \frac{1}{2}, \frac{1}{3}$.

17.3. Identify S_4 with the rotational symmetry group of a cube as in Chapter 8, and consider the action of A_4 on the set of vertices of the cube. Find the orbit and the stabilizer of each vertex.

17.4. Given an action of G on a set, show that every point of some orbit has the same stabilizer if and only if this stabilizer is a normal subgroup of G.

17.5. If G acts on X and H acts on Y prove that $G \times H$ acts on $X \times Y$ via

$$(g, h)(x, y) = (g(x), h(y)).$$

Check that the orbit of (x, y) is $G(x) \times H(y)$ and that its stabilizer is $G_x \times H_y$. We shall call this action the **product action** of $G \times H$ on $X \times Y$.

17.6. Here are four group actions on $\mathbb{R}^4$:

(a) The usual action of $GL_4(\mathbb{R})$.

(b) Identify $\mathbb{R}^4$ with $\mathbb{R}^2 \times \mathbb{R}^2$ and take the product action of $SO_2 \times SO_2$.

(c) Think of $\mathbb{R}^4$ as $\mathbb{C} \times \mathbb{C}$ and let SU_2 act in the usual way.

(d) Identify $\mathbb{R}^4$ with $\mathbb{R}^3 \times \mathbb{R}$ and take the product action of $SO_3 \times \mathbb{Z}$, where $\mathbb{Z}$ acts on $\mathbb{R}$ by addition.

Discuss the structure of the orbits and the stabilizers in each case.

17.7. If G acts on X and on Y, show that the formula $g((x, y)) = (g(x), g(y))$ defines an action of G on $X \times Y$. Check that the stabilizer of (x, y) is the intersection of G_x and G_y. Give an example which shows this action need not be transitive even if G acts transitively on both X and Y. We shall call this action the *diagonal action* of G on $X \times Y$.

17.8. Let $X = \{1, 2, 3, 4\}$ and let G be the subgroup of S_4 generated by (1234) and (24). Work out the orbits and stabilizers for the diagonal action of G on $X \times X$.

17.9. The group $C \times C$ is called the *torus*. Draw a picture of $C \times C$ to show why we give it this name. Describe the orbits of the following actions of $\mathbb{R}$ on the torus.

(a) The real number t sends (e^{ix}, e^{iy}) to $(e^{i(x+t)}, e^{iy})$.

(b) This time t sends (e^{ix}, e^{iy}) to $(e^{i(x+t)}, e^{i(y+t)})$.

(c) Finally, agree that t sends (e^{ix}, e^{iy}) to $(e^{i(x+t)}, e^{i(y+t\sqrt{2})})$.

17.10. Let x be an element of a group G. Show that the elements of G which commute with x form a subgroup of G. This subgroup is called the *centraliser* of x and written $C(x)$. Prove that the size of the conjugacy class of x is equal to the index of $C(x)$ in G. If some conjugacy class contains precisely two elements, show that G cannot be a simple group.

17.11. If n is odd show there are exactly two conjugacy classes of n-cycles in A_n each of which contains $(n - 1)!/2$ elements. When n is even, prove that the $(n - 1)$-cycles in A_n make up two conjugacy classes, each of which contains $(n - 2)!n/2$ elements.

17.12. Let p be a prime number. Show that the matrices

$$\begin{bmatrix} 1 & a & b \\ 0 & 1 & c \\ 0 & 0 & 1 \end{bmatrix}, \qquad a, b, c \in \mathbb{Z}_p$$

form a non-abelian group of order p^3.

17.13. If G is a finite group which acts transitively on X, and if H is a normal subgroup of G, show that the orbits of the induced action of H on X all have the same size.

17.14. Let H be a finite subgroup of a group G. Verify that the formula

$$(h, h')(x) = hxh'^{-1}$$

defines an action of $H \times H$ on G. Prove that H is a normal subgroup of G if and only if every orbit of this action contains precisely $|H|$ points.

CHAPTER 18

Counting Orbits

Two children, Jerome and Emily, each have a supply of cubes, a pot of red paint, and a pot of green paint. Emily decides to decorate her cubes by painting each face either red or green. Jerome plans to bisect each face with either a red or green stripe as in Figure 18.1 so that no two of his stripes meet. Who produces the largest number of differently decorated cubes?

Eventually Jerome wins 12 to 10. Children realise instinctively that rotations of the cube play a vital role. Painting the top red and all the other faces green produces the same result, from a decorative point of view, as colouring the bottom red and all the rest green. We can obtain one from the other by just turning the cube upside down, and we must agree that two coloured cubes are the same if one can be rotated into the other.

Trial and error is not the way to solve this type of problem. We are unlikely to discover all 9,099 different ways of colouring each face of a dodecahedron red, white, or blue by experiment! Instead, we analyse Emily's problem as follows. Painting each face either red or green produces an object which we call a *coloured cube*. The cube has six faces, so there are 2^6 coloured cubes in all, and they form a set which we denote by X. The rotational symmetry group of the cube acts on X, and two coloured cubes are genuinely *different* provided they do not lie in the same orbit. To solve our problem we need the number of distinct orbits of the action.

Suppose we have an action of a finite group G on a set X. Write X^g for the subset of X consisting of those points which are *left fixed* by the element g of G.

(18.1) The Counting Theorem. *The number of distinct orbits is*

$$\frac{1}{|G|} \sum_{g \in G} |X^g|$$

in other words, the average number of points left fixed by an element of G.

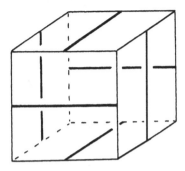

Figure 18.1

Proof. Count the collection of those ordered pairs (g, x) from $G \times X$ for which $g(x) = x$. The number of such pairs is

$$\sum_{g \in G} |X^g|. \tag{*}$$

It is also equal to

$$\sum_{x \in X} |G_x|. \tag{**}$$

Let $X_1, X_2, \ldots, X_k$ be the distinct orbits and rewrite (**) as

$$\sum_{i=1}^{k} \sum_{x \in X_i} |G_x|.$$

Points in the same orbit have conjugate stabilizers, so if $\bar{x}$ is some chosen point of X_i, we have

$$\sum_{x \in X_i} |G_x| = |X_i| \cdot |G_{\bar{x}}|$$

$$= |G(\bar{x})| \cdot |G_{\bar{x}}|$$

which is just $|G|$ by the Orbit-Stabilizer theorem. Therefore expression (**) is equal to $k|G|$. Equating (*) and (**) gives the result. □

(18.2) Theorem. *Group elements which are conjugate fix the same number of points.*

Proof. Suppose g and h are conjugate in G; say $ugu^{-1} = h$. If g fixes x, then h fixes $u(x)$ because

$$h(u(x)) = ugu^{-1}(u(x)) = ug(x) = u(x).$$

Therefore, u sends the set X^g into X^h. The same argument, with the roles of g and h reversed, shows that u^{-1} sends X^h back to X^g. This means that u provides a bijection from X^g to X^h, and the two sets of fixed points must have the same size. □

Now for our original problems.

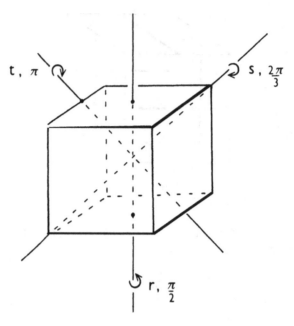

Figure 18.2

Emily's Problem. We must take an element from each conjugacy class of the rotational symmetry group of the cube and work out how many coloured cubes it leaves fixed. As representatives of the conjugacy classes we choose the rotations r, r^2, s, and t shown in Figure 18.2, together with the identity element. A coloured cube which is left fixed by r must have all four vertical faces painted the same colour because r rotates each of these to the position of its right-hand neighbour. We have a choice of two colours for the top, two for the bottom, and two for all the rest; therefore $|X^r| = 2^3$. The effect of s on the faces of the cube can be summarised by

$$\text{top} \to \text{right-hand side} \to \text{back} \to \text{top}$$

$$\text{bottom} \to \text{left-hand side} \to \text{front} \to \text{bottom}$$

giving $|X^s| = 2^2$. We leave r^2 and t to the reader, r^2 fixes 2^4 coloured cubes and t fixes 2^3. Of course the identity fixes all 2^6. The conjugacy classes of r, r^2, s, t contain six, three, eight, and six elements respectively. Therefore, the number of genuinely different coloured cubes which can be obtained by painting each face either red or green is

$$\tfrac{1}{24}\{(6 \times 2^3) + (3 \times 2^4) + (8 \times 2^2) + (6 \times 2^3) + 2^6\}$$

$$= \tfrac{1}{3}\{6 + 6 + 4 + 6 + 8\}$$

$$= 10. \qquad \qquad \qquad \qquad \Box$$

Jerome's Problem. We have lines drawn on the faces of the cube as in Figure 18.1 and plan to paint each line either red or green. This is almost the same problem as before, but not quite. The given pattern of lines is not sent into itself by every rotational symmetry of the cube, only by half of them, those which induce *even* permutations of the four leading diagonals. The group involved is A_4 rather than S_4. The conjugacy classes are represented by r^2, s, s^2, e and the number of different coloured cubes is now

$$\tfrac{1}{12}\{(3 \times 2^4) + (4 \times 2^2) + (4 \times 2^2) + 2^6\}$$
$$= \tfrac{1}{3}\{12 + 4 + 4 + 16\}$$
$$= 12. \qquad \qquad \square$$

We finish with a third problem of this type (suggested by L.M. Woodward). A 5×1 rectangular strip of paper is marked off on both sides into five unit squares. The ends of the strip are then joined with a half twist to produce a *Möbius band M*. How many different bands can result if we have three colours with which to paint the squares? There are ten squares and three colours, giving a total of 3^{10} painted bands. "Different" now means different up to the "natural" symmetry of the Möbius band. Make a model of M and run it through your fingers so that the squares move along one position. Call this symmetry r. After ten moves you get back to where you started, so r^{10} is the identity. (Note that r is *not* induced by a rotation of $\mathbb{R}^3$; it is a movement of the Möbius band in itself. This type of symmetry shows up well if the band is used as a belt drive connecting two pulleys.) There is another natural symmetry s; just turn M over as in Figure 18.3. Together r and s generate a group which is isomorphic to the *dihedral group D_{10}* and which acts on the set of painted bands in the obvious way. The conjugacy classes of D_{10} are

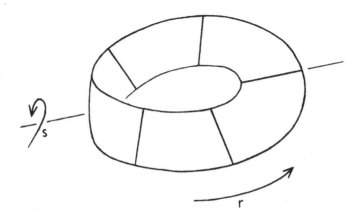

Figure 18.3

$$\{e\}, \quad \{r, r^9\}, \quad \{r^2, r^8\},$$
$$\{r^3, r^7\}, \quad \{r^4, r^6\}, \quad \{r^5\}$$
$$\{s, r^2 s, r^4 s, r^6 s, r^8 s\},$$

and

$$\{rs, r^3 s, r^5 s, r^7 s, r^9 s\}.$$

Taking the first element from each of these and working out how many painted bands it leaves fixed gives: 3^{10} for e; 3 for r; 3^2 for r^2; 3 for r^3; 3^2 for r^4; 3^5 for r^5; 3^6 for s and 3^5 for rs. (For example, s leaves two squares invariant, those which are pierced at their centre by its axis, and interchanges the other eight in pairs, so it fixes 3^6 painted bands.) By the Counting Theorem, the number of distinct painted Möbius bands is

$$\tfrac{1}{20}\{3^{10} + (2 \times 3) + (2 \times 3^2) + (2 \times 3) + (2 \times 3^2) + (1 \times 3^5)$$
$$+ (5 \times 3^6) + (5 \times 3^5)\}$$
$$= 3210.$$

The reader may well wish to have explicit formulae for the symmetries r and s. The neatest approach is as follows. Think of M as the subset

$$\{(e^{2i\theta}, \lambda e^{i\theta}) \mid -\pi < \theta \leqslant \pi, 0 \leqslant \lambda \leqslant 1\}$$

of $\mathbb{C} \times \mathbb{C}$. Then r sends $(e^{2i\theta}, \lambda e^{i\theta})$ to $(e^{2i(\theta + \pi/5)}, \lambda e^{i(\theta + \pi/5)})$, and s sends $(e^{2i\theta}, \lambda e^{i\theta})$ to $(e^{-2i\theta}, \lambda e^{-i\theta})$.

Applications of the Counting Theorem to Chemistry may be found in Section 20 of Reference [5].

EXERCISES

18.1. Each edge of a cube is painted black or white. How many different decorated cubes result?

18.2. Show there really are 9,099 essentially different ways of colouring the faces of a dodecahedron red, white, or blue.

18.3. A circular birthday cake is subdivided into eight equal wedges. In how many different ways can we distribute red and green candles so there is a candle at the centre of each piece?

18.4. Instead of painting the stripes in Figure 18.1, paint each half of the subdivided faces. In how many different ways can this be done if we have two colours available?

18.5. A bracelet is made from five beads mounted on a circular wire. How many different bracelets can we manufacture if we have red, blue, and yellow beads at our disposal?

18.6. The vertices, the midpoints of the edges, and the centroids of the faces of a regular tetrahedron T are to be labelled using three colours. Prove that there are 400,707 ways of doing this if we take into account the rotational symmetry of T.

18.7. How many different ways are there of colouring the vertices and edges of a regular hexagon using red, blue, or yellow for the edges and black or white for the vertices?

18.8. Look at our description of the Möbius band as a subset of $\mathbb{C} \times \mathbb{C}$ and find matrices in U_2 which represent the symmetries r and s.

CHAPTER 19

Finite Rotation Groups

The special orthogonal group SO_3 may be identified with the group of *rotations* of $\mathbb{R}^3$ which fix the origin (Chapter 9). If an object is positioned in $\mathbb{R}^3$ with its centre of gravity at the origin, then its rotational symmetry group "is" a subgroup of SO_3. We are familiar with several possibilities. From a right regular pyramid with an n-sided base we obtain a cyclic group of order n, while a regular plate with n sides exhibits dihedral symmetry and gives D_n. (Regular with two sides means the lens shape described in Exercise 9.12.) In addition, we have the symmetry groups of the regular solids. As we shall see, these are the *only possibilities*, provided our object has only a finite amount of symmetry. In other words a *finite* subgroup of SO_3 is either cyclic, dihedral, or isomorphic to the rotational symmetry group of one of the regular solids. We begin with a less ambitious result which deals with finite subgroups of O_2.

(19.1) Theorem. *A finite subgroup of O_2 is either cyclic or dihedral.*

Proof. Let G be a finite non-trivial subgroup of O_2. Suppose first of all that G lies inside SO_2 so that each element of G represents a *rotation* of the plane. Write A_θ for the matrix which represents rotation anticlockwise through θ about the origin, where $0 \leqslant \theta < 2\pi$, and choose $A_\varphi \in G$ so that φ is positive and as small as possible. Given $A_\theta \in G$, divide θ by φ to produce $\theta = k\varphi + \psi$ where $k \in \mathbb{Z}$ and $0 \leqslant \psi < \varphi$. Then

$$A_\theta = A_{k\varphi + \psi} = (A_\varphi)^k A_\psi \qquad \text{and} \qquad A_\psi = (A_\varphi)^{-k} A_\theta.$$

Since A_θ and A_φ both lie in G, we see that A_ψ is also in G. This gives $\psi = 0$, since otherwise we contradict our choice of φ. Therefore, G is generated by A_φ and is *cyclic*.

If G is not wholly contained inside SO_2, we set $H = G \cap SO_2$. Then H is a subgroup of G which has index 2, and by the first part H is cyclic because it is contained in SO_2. Choose a generator A for H and an element B from $G - H$. As B represents a *reflection* we have $B^2 = I$. If $A = I$, then G consists of I and B and is a *cyclic group* of order 2. Otherwise, the order of A is an integer $n \geqslant 2$. The elements of G are now

$$I, A, \ldots, A^{n-1}, B, AB, \ldots, A^{n-1}B$$

and they satisfy $A^n = I$, $B^2 = I$, $BA = A^{-1}B$. In this case the correspondence $A \to r$, $B \to s$ determines an isomorphism between G and the *dihedral group* D_n.

$\square$

(19.2) Theorem. *A finite subgroup of SO_3 is isomorphic either to a cyclic group, a dihedral group, or the rotational symmetry group of one of the regular solids.*

Proof. Let G be a finite subgroup of SO_3. Then each element of G, other than the identity, represents a rotation of $\mathbb{R}^3$ about an axis which passes through the origin. We shall argue geometrically and work with rotations rather than the corresponding matrices. The two points where the axis of a rotation $g \in G$ meets the unit sphere are called the *poles* of g (Fig. 19.1). If the axis happens to be the z-axis, we have the usual North and South poles of the sphere. These poles are the only points on the unit sphere which are left fixed by the given rotation. Let X denote the set of all poles of all elements of $G - \{e\}$. Suppose $x \in X$ and $g \in G$. Let x be a pole of the element $h \in G$. Then $(ghg^{-1})(g(x)) = g(h(x)) = g(x)$, which shows that $g(x)$ is a pole of ghg^{-1} and hence $g(x) \in X$. *Therefore, we have an action of G on X.* The idea of the proof is to apply the Counting Theorem to this action and show that X has to be a particularly nice configuration of points.

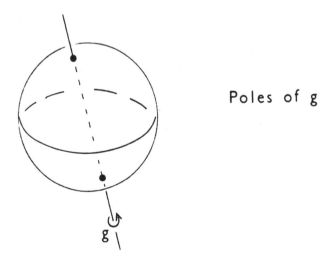

Poles of g

Figure 19.1

Let N denote the number of distinct orbits, choose a pole from each orbit, and call these poles $x_1, x_2, \ldots, x_N$. Every element of $G - \{e\}$ fixes precisely two poles, while the identity fixes them all, so the Counting Theorem gives

$$N = \frac{1}{|G|} \{2(|G| - 1) + |X|\}$$

$$= \frac{1}{|G|} \left\{ 2(|G| - 1) + \sum_{i=1}^{N} |G(x_i)| \right\}.$$

This rearranges to

$$2\left(1 - \frac{1}{|G|}\right) = N - \frac{1}{|G|} \sum_{i=1}^{N} |G(x_i)|$$

$$= N - \sum_{i=1}^{N} \frac{1}{|G_{x_i}|} \qquad\qquad (*)$$

$$= \sum_{i=1}^{N} \left(1 - \frac{1}{|G_{x_i}|}\right).$$

Assuming G is not the trivial group, the left-hand side of the above expression is greater than or equal to 1 and less than 2. But each stabilizer G_x has order at least 2 so that

$$\frac{1}{2} \leqslant 1 - \frac{1}{|G_{x_i}|} < 1$$

for $1 \leqslant i \leqslant N$. *Therefore, N is either 2 or 3.*

If $N = 2$, then $(*)$ gives $2 = |G(x_1)| + |G(x_2)|$ and there can only be two poles. These poles determine an axis L and every element of $G - \{e\}$ must be a rotation about this axis. The plane which passes through the origin and which is perpendicular to L is rotated on itself by G. Therefore, G is isomorphic to a subgroup of SO_2 and has to be *cyclic* by (19.1).

When $N = 3$ the situation is more complicated. Writing x, y, z instead of x_1, x_2, x_3 we have

$$2\left(1 - \frac{1}{|G|}\right) = 3 - \left(\frac{1}{|G_x|} + \frac{1}{|G_y|} + \frac{1}{|G_z|}\right)$$

and, therefore,

$$1 + \frac{2}{|G|} = \frac{1}{|G_x|} + \frac{1}{|G_y|} + \frac{1}{|G_z|}.$$

The sum of the three terms on the right-hand side is *greater* than 1, so there are only four possibilities:

(a) $\frac{1}{2}, \frac{1}{2}, \frac{1}{n}$ where $n \geqslant 2$;

(b) $\frac{1}{2}, \frac{1}{3}, \frac{1}{3}$;

(c) $\frac{1}{2}, \frac{1}{3}, \frac{1}{4}$;

(d) $\frac{1}{2}, \frac{1}{3}, \frac{1}{5}$.

We shall consider each of these cases in turn.

Case (a). If $|G_x| = |G_y| = |G_z| = 2$, then G is a group of order 4 in which every element other than the identity has order 2. Therefore, G is isomorphic to *Klein's group*, thought of here as a dihedral group with four elements. Let g generate G_z and remember that g preserves distance. The poles x and $g(x)$ are equidistant from z, as are y and $g(y)$. So $-z$ must be the other point in $G(z)$ and we have $g(x) = -x$, $g(y) = -y$. Therefore, the axes through x, y and z are perpendicular to one another, and the three orbits are $\{\pm x\}$, $\{\pm y\}$, $\{\pm z\}$ as in Figure 19.2.

If $|G_x| = |G_y| = 2$ and $|G_z| = n \geqslant 3$, then G is a group of order $2n$. The axis through z is fixed by every rotation in the stabilizer G_z; therefore, G_z is a cyclic group of order n. Suppose g is a minimal rotation which generates G_z. The points $x, g(x), \ldots, g^{n-1}(x)$ are all distinct. To see this, suppose $g^r(x) = g^s(x)$ where $r > s$. Then $g^{r-s}(x) = x$. But z and $-z$ are the only poles which are left fixed by g^{r-s}, and x cannot be $-z$, as $|G_x| = 2$, whereas $|G_{-z}| = |G_z| = n \geqslant 3$. Because g preserves distance, we have

$$\|x - g(x)\| = \|g(x) - g^2(x)\| = \cdots = \|g^{n-1}(x) - x\|.$$

Therefore, $x, g(x), \ldots, g^{n-1}(x)$ are the vertices of a regular n-gon P. Since G consists of $2n$ rotations each of which sends P to itself, G must be the rotational symmetry group of P. Hence, G is *dihedral*. Of course the plane of P contains the origin and is perpendicular to the axis through z. The orbits of x, y and z are shown in Figure 19.3.

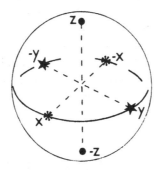

Figure 19.2

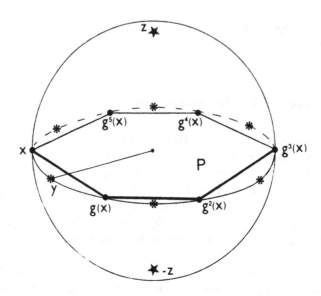

Figure 19.3

Case (b). If $|G_x| = 2$ and $|G_y| = |G_z| = 3$, then G is a group of order 12. The orbit of z consists of four points. Choose one, say u, which satisfies $0 < \|z - u\| < 2$, and choose a generator g for G_z. Then u, $g(u)$, and $g^2(u)$ are all distinct. Since g preserves distance, they are equidistant from z and lie at the corners of an equilateral triangle. Focusing our attention at u rather than z shows that z, $g(u)$, and $g^2(u)$ are equidistant from u. Therefore, z, u, $g(u)$, $g^2(u)$ are the vertices of a *regular tetrahedron*, which is sent to itself by every rotation in G. As G has the correct order (twelve), it must be the rotational symmetry group of this tetrahedron. Figure 19.4 shows the three orbits.

Case (c). If $|G_x| = 2$, $|G_y| = 3$, and $|G_z| = 4$, then G is a group of order 24. There are six points in the orbit of z. Choose one, say u, which is not z or $-z$, and let g generate G_z. Then u, $g(u)$, $g^2(u)$, and $g^3(u)$ are distinct, equidistant from z, and lie at the corners of a square. We have room for only one more point in $G(z)$, so this point must be $-z$. Equally well, $-u$ lies in $G(u) = G(z)$. This pole $-u$ is certainly not z or $-z$, and it cannot be $g(u)$ or $g^3(u)$ because $\|g(u) - u\| = \|g^3(u) - u\| < 2$. So $-u$ must be $g^2(u)$. Therefore, z, u, $g(u)$, $g^2(u)$, $g^3(u)$, $-z$ are the vertices of a *regular octahedron*, and G is its rotational symmetry group (see Fig. 19.5).

Case (d). If $|G_x| = 2$, $|G_y| = 3$, and $|G_z| = 5$, then G is a group of order 60. There are twelve points in the orbit of z. Choose two, say u and v, which satisfy

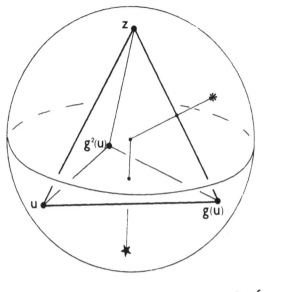

* from orbit of x
★ ——————— y

Figure 19.4

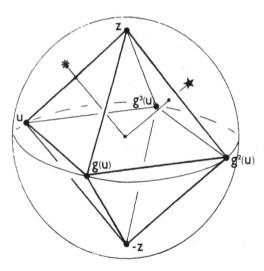

* from orbit of x
★ ——————— y

Figure 19.5

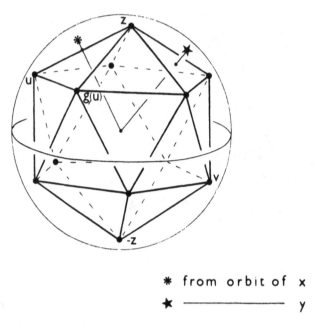

* from orbit of x

★ ——————————— y

Figure 19.6

$$0 < \|z - u\| < \|z - v\| < 2.$$

Why can we do this? If g is a minimal rotation which generates G_z, then u, $g(u)$, $g^2(u)$, $g^3(u)$, and $g^4(u)$ are all distinct, equidistant from z, and lie at the corners of a regular pentagon. Also, $v, g(v), g^2(v), g^3(v)$, and $g^4(v)$ are distinct, equidistant from z (further away than u), and form the vertices of a regular pentagon. This leaves $-z$ as the only possibility for the twelfth point of $G(z)$. *Changing our attention to u*, we see that $-u \in G(u) = G(z)$. As $-u$ lies at a distance of 2 from u, it must be one of the points $v, g(v), g^2(v), g^3(v)$, or $g^4(v)$. Relabelling if necessary, we can arrange that $-u = v$, when $-g^r(u) = g^r(v)$, $1 \leqslant r \leqslant 4$, as in Figure 19.6. Looking out from u we see eleven points, and the five which are closest to u must be equidistant from u. These are $z, g(u)$, $g^3(v), g^2(v)$, and $g^4(u)$; therefore,

$$\|u - z\| = \|u - g(u)\| = \|u - g^2(v)\|.$$

It is now easy to check that our twelve points lie at the vertices of a *regular icosahedron*, and G is the rotational symmetry group of this icosahedron. $\square$

EXERCISES

19.1. Find the poles of the rotations described in Exercise 9.8.

19.2. Glue two dodecahedra together along a pentagonal face and find the rotational symmetry group of this new solid. What is its full symmetry group?

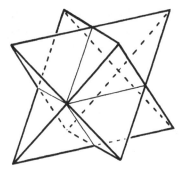

Stella octangula

Figure 19.7

19.3. A solid is made up of two regular tetrahedra arranged so that their edges bisect one another at rightangles (Fig. 19.7). Find its rotational symmetry group.

19.4. Show that the rotational symmetry group of the **cuboctahedron** (Figure 19.8) is isomorphic to S_4. A cube is made up of six pyramids, each of which has a face of the cube as its base, and the centroid of the cube as apex. By mounting these pyramids on the faces of a second cube, we obtain the **rhombic dodecahedron**. Make a model of the rhombic dodecahedron and show that its dual is the cuboctahedron.

19.5. *Finite subgroups of O_3.* Let G be a finite subgroup of O_3 and let H be its intersection with SO_3. If G contains the matrix $-I$, prove that G is

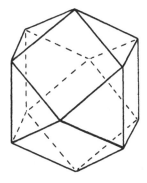

Cuboctahedron

Figure 19.8

isomorphic to $H \times \mathbb{Z}_2$. If on the other hand $-I$ does not belong to G, show that the function $A \rightarrow (\det A). A$ provides an isomorphism from G to a subgroup of SO_3. Verify that all possibilities are listed in the following table.

H	G
$\{e\}$	$\{e\}, \mathbb{Z}_2$
$\mathbb{Z}_n$	$\mathbb{Z}_n, \mathbb{Z}_n \times \mathbb{Z}_2, \mathbb{Z}_{2n}$
D_n	$D_n, D_n \times \mathbb{Z}_2, D_{2n}$
A_4	$A_4, A_4 \times \mathbb{Z}_2, S_4$
S_4	$S_4, S_4 \times \mathbb{Z}_2$
A_5	$A_5, A_5 \times \mathbb{Z}_2$

19.6. Realise each of $\mathbb{Z}_{2n}$, D_{2n}, and $A_4 \times \mathbb{Z}_2$ as the (full) symmetry group of an appropriate solid.

19.7. Find solids whose rotational symmetry groups are isomorphic to SO_2 and O_2 respectively. Find a solid whose full symmetry group is isomorphic to $O_2 \times \mathbb{Z}_2$. Show that $SO_2 \times \mathbb{Z}_2$ cannot be realised as the rotational symmetry group, or the full symmetry group, of a solid.

19.8. Prove that SO_3 does not contain a subgroup which is isomorphic to $SO_2 \times SO_2$.

CHAPTER 20

The Sylow Theorems

Let G be a finite group whose order is divisible by the prime number p. Suppose p^m is the *highest* power of p which is a factor of $|G|$ and set $k = |G|/p^m$.

(20.1) Theorem. *The group G contains at least one subgroup of order p^m.*

(20.2) Theorem. *Any two subgroups of G of order p^m are conjugate.*

(20.3) Theorem. *The number of subgroups of G of order p^m is congruent to 1 modulo p and is a factor of k.*

These results were first published by L. Sylow rather more than a century ago. The arguments presented below (the first is due to H. Wielandt, 1959) make repeated use of the Orbit-Stabilizer theorem. Each reference to this theorem will be indicated by an asterisk so as to avoid tedious repetition.

Proof of (20.1). Let X denote the collection of all subsets of G which have p^m elements and let G act on X by *left translation*, so that the group element $g \in G$ sends the subset $A \in X$ to gA. The size of X is the binomial coefficient $\binom{kp^m}{p^m}$, which is not divisible by p (see Exercise 20.14). *Hence, there must be an orbit $G(A)$ whose size is not a multiple of p.* We have $|G| = |G(A)| \cdot |G_A|$ (*), consequently $|G_A|$ is divisible by p^m. Now G_A is the stabilizer of A, so if $a \in A$ and $g \in G_A$, then $ga \in A$. This means that the whole right coset $G_A a$ is contained in A whenever $a \in A$, and $|G_A|$ cannot exceed p^m. Therefore, G_A is a subgroup of G which has order p^m. □

Proofs of (20.2) and (20.3). Let $H_1, \ldots, H_t$ denote the subgroups of G which have order p^m, and let H_1 act on the set $\{H_1, \ldots, H_t\}$ by *conjugation* so that

$h \in H_1$ sends H_j to hH_jh^{-1}. If K_j is the stabilizer of H_j, then $K_j = H_1 \cap H_j$. (So as not to interrupt the flow of our argument, we prove this as a lemma below.) In particular, $K_1 = H_1$ and the orbit of H_1 has just one element (*), namely H_1 itself. If j is not equal to 1, the order of K_j is a smaller power of p than p^m, so the size of every other orbit is a multiple of p (*). Adding up the sizes of the orbits shows that t is congruent to 1 modulo p.

Now let the *whole group G* act on $\{H_1, \ldots, H_t\}$ by conjugation. In order to prove (20.2) we must verify that this G-action is transitive. Each G-orbit is made up of various H_1-orbits. The G-orbit of H_1 certainly contains H_1, and therefore its size is congruent to 1 modulo p. Suppose now that H_r is *not* in the G-orbit of H_1, and let H_r act on $\{H_1, \ldots, H_t\}$ by conjugation. The G-orbit of H_1 is now partitioned into H_r-orbits and the size of each of these is a multiple of p (because the exceptional orbit $\{H_r\}$ is not present). This leads us to conclude that $|G(H_1)|$ is congruent to zero modulo p, which does not agree with our previous calculation. Therefore the G-orbit of H_1 must be all of $\{H_1, \ldots, H_t\}$ as required.

Since the size of an orbit is always a factor of the order of the group involved (*), we now know that t divides kp^m. But p does not divide t, so that t must be a factor of k. Our argument is now complete except for one detail.

Lemma. K_j *is the intersection of* H_1 *and* H_j.

Proof of the lemma. By definition, $K_j = \{h \in H_1 | hH_jh^{-1} = H_j\}$ and therefore $K_j \subseteq H_1$ and $H_1 \cap H_j \subseteq K_j$. We must show that K_j is contained in H_j. Certainly $K_jH_j = H_jK_j$, so K_jH_j is a *subgroup* of G (see Exercise 15.1). In addition, H_j sits inside K_jH_j as a normal subgroup and the Second Isomorphism Theorem gives

$$K_jH_j/H_j \cong K_j/(K_j \cap H_j).$$

The order of K_jH_j is therefore $|K_j| \cdot |H_j|/|K_j \cap H_j|$, which is a *power of p*. But the largest available power of p is $p^m = |H_j|$, hence $K_jH_j = H_j$, and we have $K_j \subseteq H_j$ as required. This ends the proofs of (20.1)–(20.3).   □

We shall use the Sylow theorems to help classify groups of order 12. First we make a minor though useful observation. If H is a subgroup of G, then each conjugate gHg^{-1} is also a subgroup of G and has the same order as H. Therefore, if G has no other subgroup of the same order as H, then H must be a *normal* subgroup of G.

Classification of Groups of Order 12

Let G be a group which contains twelve elements and suppose it has t subgroups of order 3. The Sylow theorems tell us that t is congruent to 1 modulo 3 and is a factor of 4. *Therefore, G has either a single subgroup of order* 3, *which must be a normal subgroup, or four conjugate subgroups of order* 3. We shall treat these two cases separately. Note that the Sylow theorems also predict either one or three subgroups of order 4.

Case 1. Suppose G contains a normal subgroup H of order 3 generated by x. Let K be a subgroup of order 4. Then K is either cyclic or isomorphic to Klein's group $\mathbb{Z}_2 \times \mathbb{Z}_2$.

(a) Assume for the moment that K is *cyclic* and choose a generator y of K. The identity is the only element which lies in both H and K, so the cosets H, Hy, Hy^2, Hy^3 are all distinct and $HK = G$. Since H is a normal subgroup, we know that yxy^{-1} belongs to H.

(i) If $yxy^{-1} = x$, then G is abelian and we have

$$G \cong H \times K \cong \mathbb{Z}_3 \times \mathbb{Z}_4 \cong \mathbb{Z}_{12}$$

by (10.1) and (10.2).

(ii) The other possibility, $yxy^{-1} = x^2$, or equivalently, $yx = x^2y$, leads to

$$y^2x = yx^2y = x^2yxy$$
$$= x^4y^2 = xy^2.$$

So x commutes with y^2 and the element $z = xy^2$ has *order* 6. We also have

$$z^3 = x^3y^6 = y^2$$

and

$$yz = yxy^2 = y^3x$$
$$= y^2x^2y = z^{-1}y.$$

The order of y is 4; therefore, y does not lie in $\langle z \rangle$ and the cosets $\langle z \rangle$, $\langle z \rangle y$ fill out G. Their elements multiply via

$$z^a z^b = z^{a+b},$$
$$z^a(z^b y) = z^{a+b}y,$$
$$(z^a y)z^b = z^{a-b}y,$$
$$(z^a y)(z^b y) = z^{a-b}y^2 = z^{a-b+3}$$

where the powers of z are read modulo 6. Associativity requires a little patience, though no real flair, and we do have a group. This group belongs to the family of *dicyclic groups* whose first member is the quaternion group Q (see Exercise 20.15).

(b) Now suppose that K is isomorphic to *Klein's group*, labelling its elements e, u, v, w, where $w = uv$ and $u^2 = v^2 = e$. Again, $H \cap K = \{e\}$, and the cosets H, Hu, Hv, Hw fill out G, giving $HK = G$. Since H is normal in G, we have

$$uxu^{-1} = x^a, \qquad vxv^{-1} = x^b, \qquad wxw^{-1} = x^{ab}$$

where each of a, b, ab is either $+1$ or -1.

(iii) If $a = b = ab = +1$, our group is *abelian* and isomorphic to $H \times K$, hence

$$G \cong \mathbb{Z}_3 \times \mathbb{Z}_2 \times \mathbb{Z}_2 \cong \mathbb{Z}_6 \times \mathbb{Z}_2.$$

(iv) Otherwise two of a, b, ab, equal -1 and the third is $+1$. Relabel u, v, w if necessary so that $a = +1$ and $b = -1$. Then u commutes with x and $z = ux$ has order 6. The elements z, v together generate G and they satisfy $z^6 = e$, $v^2 = e$, $vz = z^{-1}v$, giving a group isomorphic to the *dihedral* group of order 12.   □

Case 2. If G contains four conjugate subgroups of order 3, these subgroups use up eight elements from $G - \{e\}$, leaving room for only one subgroup K of order 4.

(c) We first show that K *cannot be cyclic*. Suppose K is cyclic, generated by the element y, and let x be an element of $G - K$. Then x has order 3 and the cosets K, Kx, Kx^2 fill out G. Since K must be normal in G, we have $xyx^{-1} \in K$. If $xyx^{-1} = y$, then G is an abelian group, contradicting our assumption that G contains four conjugate subgroups of order 3. We recognise $xyx^{-1} = y^2$ as nonsense because y and y^2 have different orders. Finally, $xyx^{-1} = y^3$ cannot hold, since it leads to

$$y = x^3yx^{-3} = y^{27} = y^3.$$

(d) The normal subgroup K of order 4 must therefore be isomorphic to *Klein's group*. Label its elements e, u, v, w as before, and let x be an element of G which has order 3. The cosets K, Kx, Kx^2 are all distinct, so u, v, and x together generate G. Conjugation by x permutes the three elements u, v, w among themselves, since K is a normal subgroup of G, and this permutation is either the identity or a 3-cycle because $x^3 = e$.

(v) We cannot obtain the identity permutation, as this leads us to an abelian group, and hence to a contradiction as above.

(vi) Relabelling as necessary, we may assume that $xux^{-1} = v$, $xvx^{-1} = w$, $xwx^{-1} = u$ when the correspondence

$$u \leftrightarrow (12)(34), \qquad v \leftrightarrow (13)(24),$$

$$x \leftrightarrow (234)$$

determines an isomorphism between G and the *alternating* group A_4.   □

We have proved that every group of order 12 is isomorphic to one of the following: the cyclic group $\mathbb{Z}_{12}$, the product $\mathbb{Z}_6 \times \mathbb{Z}_2$, the dihedral group D_6, the dicyclic group of order 12, and the alternating group A_4.

EXERCISES

Assume the order of G is kp^m, where p is prime and is not a factor of k. Then any subgroup of G which contains p^m elements will be called a *Sylow p-subgroup* of G, or just a *Sylow subgroup* if we do not need to emphasise that it belongs to the prime p.

20.1. Show that a group of order 126 must contain a normal subgroup of order 7. Prove that a group of order 1,000 cannot be a simple group.

20.2. Write down all Sylow subgroups of A_5. Are any of these subgroups normal subgroups?

20.3. If every Sylow subgroup of G is a normal subgroup, show that G is isomorphic to the product of its Sylow subgroups.

20.4. Classify the groups which have order 1,225.

20.5. Prove that a finite abelian group is isomorphic to the product of a finite number of (abelian) groups the order of each of which is a power of a prime.

In the next three exercises p and q are both prime numbers and p is greater than q.

20.6. If p is not congruent to 1 modulo q, show that every group of order pq is cyclic.

20.7. Classify the groups of order p^2q if p is not congruent to $+1$ or -1 modulo q.

20.8. Assume p^2 is not congruent to 1 modulo q, and q^2 is not congruent to 1 modulo p. Classify the groups which have order p^2q^2.

20.9. Let p be a prime factor of the order of a finite group G. If H is a normal subgroup of G whose index is not a multiple of p, show that H must contain every Sylow p-subgroup of G.

20.10. Suppose H is a Sylow subgroup of G and let J be a subgroup of G which contains H. If H is normal in J, and if J is normal in G, prove that H is normal in G.

20.11. Let H be a subgroup of G and write X for the set of left cosets of H in G. Show that the formula

$$g(xH) = gxH$$

defines an action of G on X. Prove that H is a normal subgroup of G if and only if every orbit of the induced action of H on X contains just one point.

20.12. Let G be a finite group and let p be the smallest prime which is a factor of $|G|$. Prove that a subgroup of G which has index p must be a normal subgroup of G. (You may like to use the action defined in Exercise 20.11.)

20.13. If J is a subgroup of G whose order is a power of a prime p, verify that J must be contained in a Sylow p-subgroup of G. (Take H to be a Sylow p-subgroup in Exercise 20.11 and consider the induced action of J on X.)

20.14. If p is a prime number, if k is not a multiple of p, and if $0 \leqslant x \leqslant p^m - 1$, show that $(kp^m - x)/(p^m - x)$ is not divisible by p.

20.15. Let m be an integer which is greater than or equal to 2. Begin with $4m$ elements

$$e, x, \ldots, x^{2m-1}, y, xy, \ldots, x^{2m-1}y$$

and multiply them via

$$x^a x^b = x^{a+b},$$
$$x^a(x^b y) = x^{a+b}y,$$
$$(x^a y)x^b = x^{a-b}y,$$
$$(x^a y)(x^b y) = x^{a-b+m}$$

where $0 \leqslant a, b \leqslant 2m - 1$ and the powers of x are read modulo $2m$. Check that this defines a group G which is isomorphic to the quaternion group when $m = 2$. We call G the **dicyclic group** of order $4m$.

CHAPTER 21

Finitely Generated Abelian Groups

A group is *finitely generated* if it has a finite set of generators. Finitely generated abelian groups may be classified. By this we mean we can draw up a list (albeit infinite) of "standard" examples, no two of which are isomorphic, so that if we are presented with an *arbitrary* finitely generated abelian group, it is isomorphic to one on our list.

(21.1) Theorem. *Any finitely generated abelian group is isomorphic to a direct product of cyclic groups*

$$\mathbb{Z}_{m_1} \times \mathbb{Z}_{m_2} \times \cdots \times \mathbb{Z}_{m_k} \times \mathbb{Z}^s$$

in which m_i is a factor of m_{i+1} for $1 \leqslant i \leqslant k - 1$.

We use $\mathbb{Z}^s$ as shorthand for the direct product of s copies of the additive group of integers. The number s is the **rank** of our group, and $m_1, \ldots, m_k$ are its **torsion coefficients**. Two special cases are worth listing separately.

(21.2) Corollary. *Any finite abelian group is isomorphic to a direct product of cyclic groups*

$$\mathbb{Z}_{m_1} \times \mathbb{Z}_{m_2} \times \cdots \times \mathbb{Z}_{m_k}$$

for which $m_1|m_2|\ldots|m_k$. (Here the rank s is zero.)

(21.3) Corollary. *Any finitely generated abelian group in which there are no elements of finite order is isomorphic to the direct product of a finite number of copies of $\mathbb{Z}$.* (This fits into the general case if we allow k to be zero. A group which is isomorphic to the direct product of s copies of $\mathbb{Z}$ is called a *free abelian* group of rank s.)

EXAMPLES. (i) Do not be fooled by a group such as $\mathbb{Z}_6 \times \mathbb{Z}_{10}$. At first sight we may be puzzled because 6 is not a factor of 10. But using (10.1) we have

$$\mathbb{Z}_6 \times \mathbb{Z}_{10} \cong \mathbb{Z}_2 \times \mathbb{Z}_3 \times \mathbb{Z}_{10} \cong \mathbb{Z}_2 \times \mathbb{Z}_{30},$$

which is on our list.

(ii) An abelian group of order 12 must be isomorphic to either $\mathbb{Z}_{12}$ or $\mathbb{Z}_2 \times \mathbb{Z}_6$. This bears out our calculations in Chapter 20.

(iii) The subgroup of $\mathbb{C} - \{0\}$ generated by the two elements -1 and $i/2$ is isomorphic to $\mathbb{Z}_2 \times \mathbb{Z}$.

(iv) $\mathbb{R}$ is not finitely generated (Exercise 21.12) and is not isomorphic to one of the groups on our list.

Let G be a finitely generated abelian group and let $x_1, \ldots, x_r$ be distinct elements which together generate G. If no set of $r - 1$ elements can generate G, we call $x_1, \ldots, x_r$ a *minimal* set of generators. Because our group is abelian, each $g \in G$ can be written in a particularly nice way as a word

$$g = x_1^{n_1} x_2^{n_2} \ldots x_r^{n_r} \tag{$*$}$$

where $n_1, \ldots, n_r$ are integers, simply by collecting together powers of the various generators. An expression of the form

$$e = x_1^{n_1} x_2^{n_2} \ldots x_r^{n_r} \tag{$**$}$$

is called a *relation* between our generators. If q is an integer, notice that

$$x_1 x_2^q, x_2, \ldots, x_r$$

is also a minimal set of generators for G because each word

$$x_1^{n_1} x_2^{n_2} \ldots x_r^{n_r}$$

can be rewritten in terms of these new generators as

$$(x_1 x_2^q)^{n_1} x_2^{n_2 - q n_1} x_3^{n_3} \ldots x_r^{n_r}.$$

Proof of (21.1). Suppose first of all that G has a minimal set of generators $x_1, \ldots, x_r$ for which the only relation is the trivial one obtained by setting $n_1 = n_2 = \cdots = n_r = 0$ in $(**)$. Then the expression $(*)$ for g in terms of these generators is unique and the correspondence

$$g \to (n_1, n_2, \ldots, n_r)$$

is an isomorphism between G and $\mathbb{Z}^r$.

The general case needs, as one might expect, more effort. We now have a situation in which no matter how we select a minimal set of generators for G, there is always a non-trivial relation between them. *Among all relations between all possible minimal sets of generators, there will be a smallest positive exponent, say* m_1. Suppose m_1 occurs as the exponent of x_1 in the relation

$$e = x_1^{m_1} x_2^{n_2} \dots x_r^{n_r} \qquad (***)$$

between the generators $x_1, \dots, x_r$. We claim that m_1 is a factor of n_2. For if $n_2 = qm_1 + u$ where $0 \leqslant u < m_1$, then

$$e = x_1^{m_1} x_2^{qm_1+u} x_3^{n_3} \dots x_r^{n_r}$$
$$= (x_1 x_2^q)^{m_1} x_2^u x_3^{n_3} \dots x_r^{n_r}.$$

Since $x_1 x_2^q, x_2, \dots, x_r$ is also a minimal set of generators this contradicts our choice of m_1 (as the smallest positive exponent) unless u is zero. Hence, $n_2 = qm_1$ as required. In a similar fashion, we can show that m_1 is a factor of each of $n_3, \dots, n_r$ and we set $n_i = q_i m_1$ for $3 \leqslant i \leqslant r$.

Change to the new set of generators $z_1, x_2, \dots, x_r$, where

$$z_1 = x_1 x_2^q x_3^{q_3} \dots x_r^{q_r}$$

and note that relation $(***)$ now becomes

$$e = z_1^{m_1}.$$

Our initial choice of m_1 ensures that no smaller positive power of z_1 is the identity, so m_1 is the order of z_1. Let $H = \langle z_1 \rangle$ and let G_1 be the subgroup generated by $x_2, \dots, x_r$. It is easy to check that $HG_1 = G$ and $H \cap G_1 = \{e\}$. Therefore, $G \cong H \times G_1 \cong \mathbb{Z}_{m_1} \times G_1$ by (10.2).

Now work with G_1 and carry out exactly the same procedure. Again, there are two possibilities, $G_1 \cong \mathbb{Z}^{r-1}$ and $G_1 \cong \mathbb{Z}_{m_2} \times G_2$. Thus, the original group G is either isomorphic to $\mathbb{Z}_{m_1} \times \mathbb{Z}^{r-1}$, in which case we are finished, or to $\mathbb{Z}_{m_1} \times \mathbb{Z}_{m_2} \times G_2$. In the latter case the integer m_2 occurs as the exponent of, say, y_2 in the relation

$$e = y_2^{m_2} y_3^{n_3'} \dots y_r^{n_r'}$$

between the minimal set of generators $y_2, \dots, y_r$ for G_2. Since $z_1, y_2, \dots, y_r$ is a minimal set of generators for G, and since

$$e = z_1^{m_1} y_2^{m_2} y_3^{n_3'} \dots y_r^{n_r'},$$

we see that m_1 *is a factor of* m_2. We now have all the essential ingredients for our proof, and we continue with G_2. The process eventually terminates because at each stage we reduce the number of generators by one. $\square$

To complete our classification we must show that if two finitely generated abelian groups are isomorphic then they have the same rank and the same torsion coefficients.

(21.4) Theorem. *Let* $G_1 = \mathbb{Z}_{m_1} \times \mathbb{Z}_{m_2} \times \cdots \times \mathbb{Z}_{m_k} \times \mathbb{Z}^s$ *where* $m_1 | m_2 | \dots | m_k$, *and* $G_2 = \mathbb{Z}_{n_1} \times \mathbb{Z}_{n_2} \times \cdots \times \mathbb{Z}_{n_l} \times \mathbb{Z}^t$, *where* $n_1 | n_2 | \dots | n_l$. *If* G_1 *and* G_2 *are isomorphic then* $s = t$, $k = l$ *and* $m_i = n_i$ *for* $1 \leqslant i \leqslant k$.

We shall need the following simple observation.

(21.5) Lemma. *Let m, q be positive integers. The number of integers r which satisfy $0 \leqslant r < m$ and $m|qr$ is the highest common factor of m and q.*

Proof. If d is the highest common factor of m and q set $m = m'd$, $q = q'd$ so that $\mathrm{hcf}(m', q') = 1$. As $m|qr$, we have $m'd|q'dr$ and therefore $m'|r$. This means r is one of $0, m', 2m', \ldots, (d-1)m'$. Conversely each of these integers satisfies our two hypotheses and we do indeed have d possibilities.   □

(21.6) Lemma. *If $H \cong \mathbb{Z}_m$, the number of elements $x \in H$ which satisfy $x^q = e$ is the highest common factor of m and q. If $H \cong \mathbb{Z}_{m_1} \times \mathbb{Z}_{m_2} \times \cdots \times \mathbb{Z}_{m_k}$, this number is $\mathrm{hcf}(m_1, q) . \mathrm{hcf}(m_2, q) . \cdots . \mathrm{hcf}(m_k, q)$.*

Proof. This number is preserved by an isomorphism, so the first part is just a restatement of (21.5). (Since the group operation in $\mathbb{Z}_m$ is addition modulo m, the qth power of an element r is the identity precisely when m divides qr.) For the second part, if the qth power of $(r_1, r_2, \ldots, r_k)$ is to be the identity in $\mathbb{Z}_{m_1} \times \mathbb{Z}_{m_2} \times \cdots \times \mathbb{Z}_{m_k}$, then we have $\mathrm{hcf}(m_1, q)$ choices for r_1, $\mathrm{hcf}(m_2, q)$ choices for r_2, and so on.   □

Proof of (21.4)—torsion coefficients. The elements of finite order in G_1 form the subgroup $H_1 = \mathbb{Z}_{m_1} \times \mathbb{Z}_{m_2} \times \cdots \times \mathbb{Z}_{m_k} \times \{e\}$ and those in G_2 make up $H_2 = \mathbb{Z}_{n_1} \times \mathbb{Z}_{n_2} \times \cdots \times \mathbb{Z}_{n_l} \times \{e\}$. An isomorphism between G_1 and G_2 will send elements of finite order to elements of finite order, therefore H_1 is isomorphic to H_2. Assume for the sake of argument that $k \geqslant l$ and apply (21.6) to each of H_1, H_2 with $q = m_1$. Then $\mathrm{hcf}(m_1, m_1) . \mathrm{hcf}(m_2, m_1) \ldots \mathrm{hcf}(m_k, m_1) = \mathrm{hcf}(n_1, m_1) . \mathrm{hcf}(n_2, m_1) \ldots \mathrm{hcf}(n_l, m_1)$; in other words,

$$m_1^k = \mathrm{hcf}(n_1, m_1) . \mathrm{hcf}(n_2, m_1) \ldots \mathrm{hcf}(n_l, m_1).$$

Each factor on the right-hand side is at most m_1, so we must have $l = k$ and $m_1|n_1$. Now play the same trick with $q = n_1$ to produce

$$\mathrm{hcf}(m_1, n_1) . \mathrm{hcf}(m_2, n_1) \ldots \mathrm{hcf}(m_k, n_1) = n_1^k,$$

which gives $n_1|m_1$. *At this stage we know $k = l$ and $m_1 = n_1$.*
  Apply (21.6) again to H_1 and H_2 with $q = m_2$. Then

$$m_1 m_2^{k-1} = \mathrm{hcf}(n_1, m_2) . \mathrm{hcf}(n_2, m_2) \ldots \mathrm{hcf}(n_k, m_2)$$

$$= m_1 . \mathrm{hcf}(n_2, m_2) \ldots \mathrm{hcf}(n_k, m_2)$$

which gives $m_2|n_2$. Taking $q = n_2$, we have

$$m_1 . \mathrm{hcf}(m_2, n_2) \ldots \mathrm{hcf}(m_k, n_2) = m_1 n_2^{k-1}$$

so $n_2|m_2$. *Therefore, m_2 is equal to n_2.* We leave the reader to continue this process, taking q to be m_3, then n_3, and so on.   □

(21.7) Lemma. *If $\mathbb{Z}^s$ is isomorphic to $\mathbb{Z}^t$ then $s = t$.*

Proof. Assume that s is less than or equal to t. Let $\varphi\colon \mathbb{Z}^s \to \mathbb{Z}^t$ be an isomorphism, and think of $\mathbb{Z}^t$ as the subgroup of $\mathbb{R}^t$ consisting of those points whose coordinates are all integers. Use x_i to denote the element of $\mathbb{Z}^s$ which has 1 as its ith coordinate and all other coordinates zero. If $(r_1, \ldots, r_s) \in \mathbb{Z}^s$, then

$$(r_1, \ldots, r_s) = r_1 x_1 + \cdots + r_s x_s$$

and

$$\varphi(r_1, \ldots, r_s) = r_1 \varphi(x_1) + \cdots + r_s \varphi(x_s).$$

So the image of φ lies in the subspace of $\mathbb{R}^t$ spanned by $\varphi(x_1), \ldots, \varphi(x_s)$. In order to contain $\mathbb{Z}^t$, this subspace must be *the whole of* $\mathbb{R}^t$. Therefore, $s = t$ as required. ☐

Proof of (21.4)—rank. We have a homomorphism from G_1 to $\mathbb{Z}^s$ given by

$$(r_1, \ldots, r_k, r_{k+1}, \ldots, r_{k+s}) \to (r_{k+1}, \ldots, r_{k+s}).$$

It is surjective and its kernel is precisely H_1, so G_1/H_1 is isomorphic to $\mathbb{Z}^s$ by the First Isomorphism Theorem. Similarly, G_2/H_2 is isomorphic to $\mathbb{Z}^t$. As we explained earlier, an isomorphism between G_1 and G_2 sends H_1 to H_2. It will therefore induce an isomorphism from the quotient group G_1/H_1 to G_2/H_2, consequently $s = t$ by (21.7). ☐

Our proof of (21.1) is efficient, but not a great deal of help in recognising an abelian group from a given set of generators and relations. We shall show how to do this in a systematic way in the next chapter. The method uses matrix row and column operations, so we shall change to additive notation, writing $x + y$ for the group operation, 0 for the identity element, and $-x$ for the inverse of x. The standard form of (21.1) is then referred to as the abelian group determined by $k + s$ *generators* $x_1, \ldots, x_{k+s}$, which satisfy the *relations*

$$m_1 x_1 = 0, \ldots, m_k x_k = 0.$$

It is isomorphic to the quotient group A/N where A is the free abelian group whose elements are linear combinations

$$n_1 x_1 + \cdots + n_{k+s} x_{k+s}$$

with integer coefficients, and N is the subgroup of A generated by $m_1 x_1, \ldots, m_k x_k$.

EXERCISES

21.1. Find the torsion coefficients of each of the following.
  (a) $\mathbb{Z}_{10} \times \mathbb{Z}_{15} \times \mathbb{Z}_{20}$ (b) $\mathbb{Z}_{28} \times \mathbb{Z}_{42}$
  (c) $\mathbb{Z}_9 \times \mathbb{Z}_{14} \times \mathbb{Z}_6 \times \mathbb{Z}_{16}$

21.2. Let G be an abelian group of order 100. Show that G must contain an element of order 10. What are the torsion coefficients of G if no element of G has order greater than 10?

21.3. If the order of a finite abelian group is not divisible by a square, show that the group must be cyclic.

21.4. Classify the abelian groups of order 81, 144, and 216.

21.5. Let p be a prime number. An abelian group has order p^n and contains $p - 1$ elements of order p. Show that the group is cyclic.

21.6. If G, A, B are finite abelian groups, and if $G \times A$ is isomorphic to $G \times B$, prove that A is isomorphic to B.

21.7. Let G be a finite abelian group of order 360 which does not contain any elements of order 12 or 18. Find the torsion coefficients of G. How many elements of order 6 does G contain?

21.8. Prove that a finitely generated non-trivial subgroup of $\mathbb{R} - \{0\}$ must be isomorphic to $\mathbb{Z}_2$ or to $\mathbb{Z}^s$ or $\mathbb{Z}_2 \times \mathbb{Z}^s$ for some positive integer s.

21.9. Let G be a finite abelian group and write $\#(q)$ for the number of elements x of G which satisfy $x^q = e$. Find the torsion coefficients of G when $\#(2) = 16$, $\#(4) = 32$, $\#(3) = 9$, $\#(9) = 81$ and $x^{36} = e$ for all $x \in G$.

21.10. As for the previous exercise, but this time $\#(2) = 4$, $\#(3) = 3$, $\#(5) = 125$ and $x^{30} = e$ for all $x \in G$.

21.11. Abelianise each of
(a) $Q \times S_4$ (b) $D_{12} \times A_4$
(c) $G \times Z_{10}$, where G is the dicyclic group of order 12
and write down the torsion coefficients of the resulting abelian groups.

21.12. Prove that $\mathbb{R}$ cannot be generated by a finite number of elements by showing that every finitely generated group is *countable*.

21.13. Let G be an abelian group and use additive notation. Call G a **divisible group** if given $x \in G$ and a positive integer m we can always find an element y of G such that $my = x$. For example, $\mathbb{Q}$, $\mathbb{R}$, $\mathbb{C}$, and C are all divisible groups. Show that $\mathbb{Z}$ and $\mathbb{Q}^{pos}$ are not divisible. Prove that a non-trivial divisible group cannot be finitely generated.

21.14. Show that $\mathbb{Q}^{pos}$ contains a free abelian subgroup of rank s for arbitrarily large values of s.

CHAPTER 22

Row and Column Operations

Armed with generators and relations for a finitely generated abelian group, we would like to be able to recognise the group in the canonical form provided by (21.1). An effective procedure for doing this is given below. We shall use additive notation throughout, and we begin with an example.

EXAMPLE (i). Let G be the abelian group determined by the generators x, y, z and the relations

$$3x + 5y - 3z = 0 \qquad (R_1),$$

$$4x + 2y = 0 \qquad (R_2).$$

That is to say, G is isomorphic to the quotient A/N where A is the free abelian group whose elements are linear combinations $ax + by + cz$ with integer coefficients and N is the subgroup of A generated by $3x + 5y - 3z$ and $4x + 2y$. Subtract R_1 from R_2 to give

$$x - 3y + 3z = 0 \qquad (R_3)$$

and then subtract three times R_3 from R_1 so that

$$14y - 12z = 0 \qquad (R_4).$$

If $u = 3x + 5y - 3z$ and $v = 4x + 2y$, all we have done so far is to change our generators *for N* from u, v to

$$u, \quad v - u = x - 3y + 3z$$

and then to

$$v - u, \quad u - 3(v - u) = 14y - 12z.$$

Therefore, we have the same group G whether we use the relations R_1 and R_2 or R_3 and R_4. Rewrite R_4 as

$$14y - 12z = 2y + 12(y - z)$$
$$= 2(y + 6(y - z)) = 0$$

and set

$$x' = x - 3y + 3z, \qquad y' = y + 6(y - z), \qquad z' = y - z.$$

Then x', y', z' is a new set of generators *for G* as we see if we break down the change from x, y, z to x', y', z' into several stages, viz.,

$$x, \quad y, \quad z$$
$$x - 3y, \quad y, \quad z$$
$$x - 3y + 3z, \quad y \quad z$$
$$x - 3y + 3z, \quad y, \quad y - z$$
$$x - 3y + 3z, \quad y + 6(y - z), \quad y - z$$

The relations R_3, R_4 now simplify to $x' = 0$ and $2y' = 0$, respectively. So x' makes no contribution and G is generated by an element y' of order 2 together with an element z' which has infinite order. We conclude that G must be isomorphic to $\mathbb{Z}_2 \times \mathbb{Z}$.

All of these manipulations can be conveniently summarised using the *coefficient matrix* of the original relations R_1 and R_2. The corresponding steps are as follows.

$$\begin{bmatrix} 3 & 5 & -3 \\ 4 & 2 & 0 \end{bmatrix} \xrightarrow[\text{from row 2}]{\text{take row 1}} \begin{bmatrix} 3 & 5 & -3 \\ 1 & -3 & 3 \end{bmatrix} \xrightarrow[\text{rows 1 and 2}]{\text{interchange}} \begin{bmatrix} 1 & -3 & 3 \\ 3 & 5 & -3 \end{bmatrix}$$

$$\xrightarrow[\text{from row 2}]{\text{take 3 (row 1)}} \begin{bmatrix} 1 & -3 & 3 \\ 0 & 14 & -12 \end{bmatrix} \xrightarrow[\text{to col 2}]{\text{add 3 (col 1)}} \begin{bmatrix} 1 & 0 & 3 \\ 0 & 14 & -12 \end{bmatrix}$$

$$\xrightarrow[\text{3 (col 1) from col 3}]{\text{subtract}} \begin{bmatrix} 1 & 0 & 0 \\ 0 & 14 & -12 \end{bmatrix} \xrightarrow[\text{to col 2}]{\text{add col 3}} \begin{bmatrix} 1 & 0 & 0 \\ 0 & 2 & -12 \end{bmatrix}$$

$$\xrightarrow[\text{to col 3}]{\text{add 6 (col 2)}} \begin{bmatrix} 1 & 0 & 0 \\ 0 & 2 & 0 \end{bmatrix}.$$

This final matrix represents our new relations $x' = 0$, $2y' = 0$, which in turn allow us to read off the canonical form $\mathbb{Z}_2 \times \mathbb{Z}$ for G.

A general procedure has now evolved from this simple example. If G is the abelian group determined by generators $x_1, x_2, \ldots, x_n$ and relations

$$a_{11}x_1 + a_{12}x_2 + \cdots + a_{1n}x_n = 0$$
$$a_{21}x_1 + a_{22}x_2 + \cdots + a_{2n}x_n = 0$$
$$\vdots$$
$$a_{m1}x_1 + a_{m2}x_2 + \cdots + a_{mn}x_n = 0$$

we simplify the coefficient matrix $A = (a_{ij})$ using the following operations.

(I) Interchange two rows or two columns.

(II) Multiply all the entries of a row or a column by -1.

(III) Add an integer multiple of one row to another row, or an integer multiple of one column to another column.

The operations on rows alter the relations, those on columns alter the generators, *while keeping the same group G throughout.* We hope for a new set of generators and relations which present our group in its canonical form. These are guaranteed by the next result.

(22.1) Theorem. *Given an $m \times n$ matrix A whose entries are integers, there is a finite sequence of operations of type I, II, III that converts A into a diagonal matrix D for which $d_{ii} \geq 0$, $1 \leq i \leq k$, and $d_{11} | d_{22} | \ldots | d_{kk}$, where $k = \min(m, n)$.*

In terms of suitable new generators $x'_1, \ldots, x'_n$ the matrix D represents the relations $d_{11} x'_1 = 0, \ldots, d_{kk} x'_k = 0$. Abbreviate d_{ii} to d_i and suppose $d_1 \geq 2$, $d_k \neq 0$. Then the canonical form of G is

$$\mathbb{Z}_{d_1} \times \mathbb{Z}_{d_2} \times \cdots \times \mathbb{Z}_{d_k} \times \mathbb{Z}^{n-k}.$$

If, say, $d_1 = \cdots = d_s = 1$ and $d_{t+1} = \cdots = d_k = 0$, then the generators $x'_1, \ldots, x'_s$ are redundant, whereas $x'_{t+1}, \ldots, x'_k$ have infinite order. Therefore, in this case the standard form becomes

$$\mathbb{Z}_{d_{s+1}} \times \mathbb{Z}_{d_{s+2}} \times \cdots \times \mathbb{Z}_{d_t} \times \mathbb{Z}^{n-t}.$$

Outline Proof for **(22.1).** We say "outline" because the idea, though simple, is very easily lost if we become too formal. If A is the zero matrix, there is nothing to do. Otherwise, A has at least one non-zero entry and using operations of type I and II we can produce a matrix (which we still call A) whose leading entry a_{11} is *positive.* Look along the first row. If we see an entry a_{1j} that is not a multiple of a_{11}, we divide a_{1j} by a_{11} to give $a_{1j} = qa_{11} + u$, where $0 < u < a_{11}$. Subtract q times column 1 from column j (operation III), then interchange the new jth column with the first column (operation I). The resulting matrix has u as leading entry and we ask whether u now divides every other entry in the first row. If the answer is no, we repeat our procedure and produce a matrix whose leading entry v is positive and less than u. This process cannot go on indefinitely because the *decreasing* sequence a_{11}, u, v ... of positive integers must terminate. Therefore, a finite number of steps brings us to a matrix B whose leading entry divides every other entry of the first row. Using operations of type III to subtract appropriate multiples of the first column of B from the other columns, we reduce all elements of the first row, other than b_{11}, to zero. Now focus attention on the first column and proceed in exactly the same way, using row operations in place of column operations, until we reach a matrix of the form

$$\begin{bmatrix} c_{11} & 0 & \cdots & 0 \\ 0 & & & \\ \vdots & & C_1 & \\ 0 & & & \end{bmatrix} \qquad (*).$$

At this point we may be tempted to work with the smaller matrix C_1, carrying out the above procedure until we have

$$\begin{bmatrix} c_{11} & 0 & 0 & \cdots & 0 \\ 0 & c_{22} & 0 & \cdots & 0 \\ 0 & 0 & & & \\ \vdots & \vdots & & C_2 & \\ 0 & 0 & & & \end{bmatrix} \qquad (**)$$

and so on. This does give a diagonal matrix, but unfortunately its diagonal elements may not successively divide one another. So instead we ask whether every entry of C_1 is a multiple of c_{11}. If c_{11} does not divide c_{ij}, we add row i of $(*)$ to row 1 and *start the whole process again from the very beginning*. Quite depressing until we realise that this will lead to a new version of $(*)$ with a smaller positive leading entry than c_{11}. Repeat the procedure (again it must terminate after a finite number of steps) until we reach a matrix of type $(*)$ whose leading entry does divide every other entry. This leading entry is our first genuine diagonal term d_{11}. We can now begin to simplify the smaller matrix which corresponds to C_1, confident that we can reach

$$\begin{bmatrix} d_{11} & 0 & 0 & \cdots & 0 \\ 0 & d_{22} & 0 & \cdots & 0 \\ 0 & 0 & & & \\ \vdots & \vdots & & C_3 & \\ 0 & 0 & & & \end{bmatrix}$$

where $d_{11}|d_{22}$ and every element of C_3 is a multiple of d_{22}. The whole process finally comes to a halt at the required matrix D. □

EXAMPLE (ii). Let G be the abelian group determined by generators x, y and relations $2x = 0$, $3y = 0$. We immediately recognise $\mathbb{Z}_2 \times \mathbb{Z}_3$, whose standard form is $\mathbb{Z}_6$. The above procedure gives

$$\begin{bmatrix} 2 & 0 \\ 0 & 3 \end{bmatrix} \xrightarrow[\text{to row 1}]{\text{add row 2}} \begin{bmatrix} 2 & 3 \\ 0 & 3 \end{bmatrix} \xrightarrow[\text{col 1 from col 2}]{\text{subtract}} \begin{bmatrix} 2 & 1 \\ 0 & 3 \end{bmatrix}$$

$$\xrightarrow[\text{cols 1 and 2}]{\text{interchange}} \begin{bmatrix} 1 & 2 \\ 3 & 0 \end{bmatrix} \xrightarrow[\text{2 (col 1) from col 2}]{\text{subtract}} \begin{bmatrix} 1 & 0 \\ 3 & -6 \end{bmatrix}$$

$$\xrightarrow[\text{3 (row 1) from row 2}]{\text{subtract}} \begin{bmatrix} 1 & 0 \\ 0 & -6 \end{bmatrix} \xrightarrow[\text{col 2 by } -1]{\text{multiply}} \begin{bmatrix} 1 & 0 \\ 0 & 6 \end{bmatrix}$$

The new presentation has generators x', y' subject to $x' = 0$, $6y' = 0$, and we do indeed have $\mathbb{Z}_6$.

EXAMPLE (iii). Let G be the abelian group determined by generators x_1, x_2, x_3, x_4, x_5 and relations

$$x_1 - 5x_2 + 10x_4 - 15x_5 = 0$$
$$4x_2 - 8x_4 + 12x_5 = 0$$
$$3x_1 - 3x_2 - 2x_3 + 6x_4 - 9x_5 = 0$$
$$x_1 - x_2 + 2x_4 - 3x_5 = 0$$

We indicate a reduction of the coefficient matrix below. There is no need to follow the procedure of (22.1) exactly; short cuts may be taken. Here, for example, we can immediately remove columns 4 and 5, as they are both integer multiples of column 2. To save space we allow each arrow to represent several row or column operations.

$$\begin{bmatrix} 1 & -5 & 0 & 10 & -15 \\ 0 & 4 & 0 & -8 & 12 \\ 3 & -3 & -2 & 6 & -9 \\ 1 & -1 & 0 & 2 & -3 \end{bmatrix} \rightarrow \begin{bmatrix} 1 & -5 & 0 & 0 & 0 \\ 0 & 4 & 0 & 0 & 0 \\ 3 & -3 & -2 & 0 & 0 \\ 1 & -1 & 0 & 0 & 0 \end{bmatrix}$$

$$\rightarrow \begin{bmatrix} 1 & -4 & 0 & 0 & 0 \\ 0 & 4 & 0 & 0 & 0 \\ 3 & 0 & 2 & 0 & 0 \\ 1 & 0 & 0 & 0 & 0 \end{bmatrix}$$

$$\rightarrow \begin{bmatrix} 0 & -4 & 0 & 0 & 0 \\ 0 & 4 & 0 & 0 & 0 \\ 0 & 0 & 2 & 0 & 0 \\ 1 & 0 & 0 & 0 & 0 \end{bmatrix}$$

$$\rightarrow \begin{bmatrix} 1 & 0 & 0 & 0 & 0 \\ 0 & 2 & 0 & 0 & 0 \\ 0 & 0 & 4 & 0 & 0 \\ 0 & 0 & 0 & 0 & 0 \end{bmatrix}$$

Therefore, G is isomorphic to $\mathbb{Z}_2 \times \mathbb{Z}_4 \times \mathbb{Z} \times \mathbb{Z}$. (Here $m = k = 4$, $n = 5$, $s = 1$, and $t = 3$.)

EXERCISES

22.1. Find the rank and the torsion coefficients of the abelian group determined by generators x_1, x_2, x_3, x_4 and relations

$$9x_1 + 6x_2 + 5x_3 + 4x_4 = 0$$
$$6x_1 + 5x_2 - 3x_3 + 11x_4 = 0$$
$$3x_1 + 2x_2 - x_3 + 4x_4 = 0$$

22.2. An abelian group is determined by five generators and four relations, the coefficient matrix of the relations being

$$\begin{bmatrix} 3 & -4 & 5 & 3 & 7 \\ 3 & 2 & -1 & -3 & 1 \\ 8 & 2 & -2 & -8 & 2 \\ 11 & -8 & 9 & -5 & 9 \end{bmatrix}$$

Show that the group is isomorphic to $\mathbb{Z}_2 \times \mathbb{Z}_6 \times \mathbb{Z}_6 \times \mathbb{Z}$.

22.3 Let G be the abelian group which is determined by generators x_1, x_2, x_3, x_4 and relations

$$4x_1 + 2x_2 + 3x_3 + 7x_4 = 0$$

$$5x_1 + x_2 - x_3 + 12x_4 = 0$$

$$2x_1 + 4x_2 + 11x_3 - 3x_4 = 0$$

Prove that G is a free abelian group and find its rank.

22.4. How do integer row and column operations affect the determinant of a square matrix?

22.5. An abelian group G is determined by generators x, y, z and relations

$$3x + 2y + 4z = 0$$

$$6x + y + 7z = 0$$

$$2x + 3y + 6z = 0$$

Calculate the determinant of the coefficient matrix of these relations. Does the determinant give you any information about G? Find the rank and the torsion coefficients of G.

22.6. You are given n generators and n relations which determine an abelian group G. What can you say about G if the coefficient matrix of the relations has non-zero determinant? What happens if the determinant of this matrix is $+1$ or -1?

CHAPTER 23

Automorphisms

An **automorphism** of a group G is an isomorphism from G to G. The set of all automorphisms forms a group under composition of functions which is called the **automorphism group** of G and written $\text{Aut}(G)$.

EXAMPLES. (i) An automorphism θ of $\mathbb{Z}$ must send 1 to an integer which generates $\mathbb{Z}$, therefore $\theta(1) = \pm 1$. If $\theta(1) = 1$, we have the identity automorphism. Otherwise, $\theta(1) = -1$ and θ sends each integer n to $-n$. We see immediately that $\text{Aut}(\mathbb{Z})$ is isomorphic to $\mathbb{Z}_2$.

(ii) The correspondence $x \to x^{-1}$ determines an automorphism of any *abelian* group.

(iii) Suppose G is $\mathbb{Z}_2 \times \mathbb{Z}_2$. An automorphism permutes the three non-identity elements, and one easily checks that any such permutation, when completed by sending e to e, is an automorphism of $\mathbb{Z}_2 \times \mathbb{Z}_2$. Therefore, $\text{Aut}(\mathbb{Z}_2 \times \mathbb{Z}_2)$ is isomorphic to S_3.

(iv) $\text{Aut}(\mathbb{Z}_n)$ is isomorphic to the group R_n introduced in Chapter 11. The elements of R_n are the positive integers less than n which are relatively prime to n, and the group operation is multiplication modulo n. Suppose θ is an automorphism of $\mathbb{Z}_n$, then $\theta(1)$ generates $\mathbb{Z}_n$ and consequently the highest common factor of $\theta(1)$ and n must be 1. The correspondence $\theta \to \theta(1)$ is an isomorphism between $\text{Aut}(\mathbb{Z}_n)$ and R_n. We shall work out the structure of the finite abelian group R_n in Exercise 23.4.

(v) An automorphism preserves the order of each element, so an automorphism of S_3 has to permute the transpositions (12), (13), (23) among themselves. Conversely, every permutation of (12), (13), (23) determines an automorphism of S_3. Therefore $\text{Aut}(S_3)$ is isomorphic to S_3.

(vi) Let θ be an automorphism of $\mathbb{Z} \times \mathbb{Z}$. Suppose θ sends $(1,0)$ to (a,b) and $(0,1)$ to (c,d), then

$$\theta(m,n) = m\theta(1,0) + n\theta(0,1)$$
$$= (ma + nc, mb + nd)$$
$$= (m,n)\begin{bmatrix} a & b \\ c & d \end{bmatrix}$$

so that θ is represented by a 2×2 matrix with integer entries. The determinant $ad - bc$ is equal to $+1$ or -1 because θ is invertible and its inverse must also be represented by a matrix whose entries are integers. $\text{Aut}(\mathbb{Z} \times \mathbb{Z})$ is isomorphic to the group $GL_2(\mathbb{Z})$ of 2×2 matrices, which have integer entries and determinant ± 1.

(vii) Conjugation by a fixed element g of G gives a particular type of automorphism $x \to gxg^{-1}$ called an **inner automorphism**. The inner automorphisms form a normal subgroup $\text{Inn}(G)$ of $\text{Aut}(G)$. If G is abelian, only the identity automorphism is an inner automorphism.

(23.1) Theorem. $\text{Inn}(G)$ *is isomorphic to the quotient group* $G/Z(G)$.

Proof. The function from G to $\text{Aut}(G)$ which sends each g to the inner automorphism $x \to gxg^{-1}$ is a homomorphism. Its image consists of the inner automorphisms, and its kernel is

$$\{g \in G | x = gxg^{-1}, \forall x \in G\}$$
$$= \{g \in G | xg = gx, \forall x \in G\}$$
$$= Z(G).$$

The result now follows from the First Isomorphism Theorem.   □

(23.2) Theorem. *If p, q are primes which satisfy $p > q$ and $q \nmid (p-1)$, then every group of order pq is cyclic.*

Proof. Either apply the Sylow theorems (see Exercise 20.6), or proceed as follows. Let G be a group whose order is pq. Choose an element x of order p, an element y of order q, and set $H = \langle x \rangle$. We first show that H is a normal subgroup of G. The set of left cosets of H in G has q members and H acts on this set by left translation; that is to say, $h \in H$ sends the coset gH to hgH. The size of each orbit is at most q and must be a factor of $|H| = p$. Therefore, every orbit contains just one coset, in other words $hgH = gH$ for all $g \in G$ and all $h \in H$. Rewriting this as

$$g^{-1}hg \in H, \qquad \forall g \in G, h \in H$$

we see at once that H is normal in G.

The element y gives an automorphism $h \to yhy^{-1}$ of H whose order is a factor of q (because $y^q = e$) and a factor of $p - 1$ (by Lagrange because $|\text{Aut}(H)| = |\text{Aut}(\mathbb{Z}_p)| = p - 1$). As $q \nmid (p - 1)$, this order is 1 and we have the identity automorphism. Therefore, $yh = hy$ for all $h \in H$. If $K = \langle y \rangle$, then $HK = G$, $H \cap K = \{e\}$ and $hk = kh$ whenever $h \in H$, $k \in K$, consequently

$$G \cong H \times K \cong \mathbb{Z}_p \times \mathbb{Z}_q \cong \mathbb{Z}_{pq}$$

by (10.2).   □

Suppose we are given groups H, J and a homomorphism $\varphi \colon J \to \text{Aut}(H)$. We shall construct a new group $H \times_\varphi J$ called the **semidirect product** of H and J determined by φ as follows. Its elements are ordered pairs (x, y) where $x \in H$, $y \in J$, and multiplication is defined by

$$(x, y)(x', y') = (x \cdot \varphi(y)(x'), y \cdot y').$$

The first coordinate of this product is obtained by applying the automorphism $\varphi(y)$ to x', then multiplying the result on the left by x. For emphasis multiplication in both H and J has been denoted by a dot. The identity element is (e_H, e_J) and the inverse of (x, y) is

$(\varphi(y)^{-1}(x^{-1}), y^{-1})$. Associativity follows from

$$[(x, y)(x', y')](x'', y'') = (x \cdot \varphi(y)(x'), y \cdot y')(x'', y'')$$

$$= (x \cdot \varphi(y)(x') \cdot \varphi(y \cdot y')(x''), y \cdot y' \cdot y'')$$

$$= (x \cdot \varphi(y)(x' \cdot \varphi(y')(x'')), y \cdot y' \cdot y'')$$

$$\text{because } \varphi \text{ is a homomorphism}$$

$$= (x, y)(x' \cdot \varphi(y')(x''), y' \cdot y'')$$

$$= (x, y)[(x', y')(x'', y'')].$$

The function $(x, y) \to y$ is a homomorphism from $H \times_\varphi J$ onto J, whose kernel $\{(x, e_J) | x \in H\}$ is isomorphic to H. So we have a copy of H which sits inside $H \times_\varphi J$ as a *normal* subgroup. There is also a copy of J inside the semi-direct product namely $\{(e_H, y) | y \in J\}$, though this subgroup is not necessarily normal.

If φ sends every element of J to the identity automorphism of H, we recapture the direct product $H \times J$ introduced in Section 10. The next result generalises (10.2).

(23.3) Theorem. *Let H, J be subgroups of a group G. If H is a normal subgroup, if $HJ = G$ and $H \cap J = \{e\}$, then G is isomorphic to the semidirect product $H \times_\varphi J$, where $\varphi \colon J \to \text{Aut}(H)$ is the homomorphism defined by $\varphi(y)(x) = yxy^{-1}$ for all $x \in H$, $y \in J$.*

Proof. Define $\psi \colon H \times_\varphi J \to G$ by $\psi(x, y) = xy$. Then ψ is a homomorphism because

$$\psi[(x,y)(x',y')] = \psi(x \cdot \varphi(y)(x'), y \cdot y')$$
$$= \psi(xyx'y^{-1}, y \cdot y')$$
$$= xyx'y^{-1}yy'$$
$$= xyx'y'$$
$$= \psi(x,y)\psi(x',y').$$

The image of ψ is all of G because $G = HJ$, so that every element of G may be written in the form xy for some $x \in H$, $y \in J$. If (x, y) lies in the kernel of ψ, then $xy = e$, giving $x = y^{-1}$. Therefore, x and y both belong to $H \cap J = \{e\}$, and (x, y) is the identity element of $H \times_\varphi J$. By (16.3), ψ is an isomorphism.

$\square$

EXAMPLES. (i) In S_3 the subgroups $H = \langle(123)\rangle$ and $J = \langle(12)\rangle$ satisfy the hypotheses of (23.3). Therefore, S_3 is isomorphic to the semidirect product $\mathbb{Z}_3 \times_\varphi \mathbb{Z}_2$, where φ sends the generator of $\mathbb{Z}_2$ to the non-trivial automorphism of $\mathbb{Z}_3$.

(ii) The isometries (distance preserving transformations) of the plane form a group E_2 called the *Euclidean group*. Let T denote the subgroup of E_2 which consists of all the translations, and let O be the subgroup of orthogonal transformations f_A, $A \in O_2$. In the next chapter we shall check that T is a normal subgroup of E_2, $TO = E_2$, and $T \cap O = \{I\}$. Therefore, E_2 is the semidirect product $T \times_\varphi O$, where $\varphi: O \to \text{Aut}(T)$ is induced by conjugation.

EXERCISES

23.1. Find the automorphism groups of D_4 and D_5.

23.2. Work out $\text{Aut}(Q)$ and $\text{Inn}(Q)$.

23.3. If $Z(G) = \{e\}$, prove that the centre of the automorphism group of G contains only the identity automorphism.

23.4. Let m be a positive integer. Show that $R(2^m)$ is isomorphic to $\mathbb{Z}_2 \times \mathbb{Z}_{2^{m-2}}$ for $m \geqslant 3$, and that $R(p^m)$ is cyclic of order $p^{m-1}(p-1)$ when p is an odd prime. Exercise 11.6 now allows you to work out the torsion coefficients of $R(n)$ for any positive integer n.

23.5. A subgroup H of G is called a **characteristic subgroup** of G if H is sent to itself by every automorphism of G. Show that $Z(G)$ and $[G, G]$ are both characteristic subgroups of G.

23.6. Normal subgroups are precisely those which are left invariant by all inner automorphisms. Supply a group G and a normal subgroup of G which is not a characteristic subgroup.

23.7. Let G and H be finite groups whose orders are relatively prime. Show that $G \times \{e\}$ and $\{e\} \times H$ are both characteristic subgroups of $G \times H$. Prove that $\mathrm{Aut}(G \times H)$ is isomorphic to $\mathrm{Aut}(G) \times \mathrm{Aut}(H)$.

A group of the form $H \times_\varphi J$ will be called a semidirect product *of H by J*.

23.8. Prove that the dicyclic group of order 12 is isomorphic to a semidirect product of $\mathbb{Z}_3$ by $\mathbb{Z}_4$

23.9. Show that Q is not isomorphic to a semidirect product of two non-trivial groups.

23.10. Let G be an abelian group and write $G \,\tilde{\times}\, \mathbb{Z}_2$ for the semidirect product $G \times_\varphi \mathbb{Z}_2$, where $\varphi : \mathbb{Z}_2 \to \mathrm{Aut}(G)$ takes 1 to the automorphism $x \to x^{-1}$ of G. Prove that $\mathbb{Z}_n \,\tilde{\times}\, \mathbb{Z}_2$ is isomorphic to D_n, that $\mathbb{Z} \,\tilde{\times}\, \mathbb{Z}_2$ is isomorphic to D_∞, and that $SO_2 \,\tilde{\times}\, \mathbb{Z}_2$ is isomorphic to O_2.

23.11. Use the matrix

$$\begin{bmatrix} 1 & 0 & 0 \\ 0 & 1 & 0 \\ 0 & 0 & -1 \end{bmatrix}$$

and the construction of Theorem 23.3 to express O_3 as a semidirect product of SO_3 by $\mathbb{Z}_2$. Find an isomorphism between this semidirect product and $SO_3 \times \mathbb{Z}_2$.

23.12. Let G be the group of distance preserving transformations of $\mathbb{R}^2$ which is generated by $(x, y) \to (x + 1, y)$ and $(x, y) \to (-x, y + 1)$. Prove that G is isomorphic to the semidirect product $\mathbb{Z} \times_\varphi \mathbb{Z}$ where φ sends 1 to the non-trivial automorphism of $\mathbb{Z}$.

CHAPTER 24

The Euclidean Group

The *isometries* of the plane form a group under composition of functions, the so called *Euclidean group* E_2. A function $g: \mathbb{R}^2 \to \mathbb{R}^2$ belongs to E_2, provided it preserves distance; that is to say

$$\|g(\mathbf{x}) - g(\mathbf{y})\| = \|\mathbf{x} - \mathbf{y}\|$$

for every pair of points $\mathbf{x}$, $\mathbf{y}$ in $\mathbb{R}^2$. If $g, h \in E_2$, we have

$$\|g(h(\mathbf{x})) - g(h(\mathbf{y}))\| = \|h(\mathbf{x}) - h(\mathbf{y})\|$$

because g is an isometry

$$= \|\mathbf{x} - \mathbf{y}\|$$

because h is an isometry;

therefore, $gh \in E_2$. Composition of functions is associative, and the identity transformation of the plane acts as identity element. Finally, each $g \in E_2$ is a bijection and satisfies

$$\|g^{-1}(\mathbf{x}) - g^{-1}(\mathbf{y})\| = \|g(g^{-1}(\mathbf{x})) - g(g^{-1}(\mathbf{y}))\|$$

because g is an isometry

$$= \|\mathbf{x} - \mathbf{y}\|,$$

so $g^{-1} \in E_2$ and we do indeed have a group.

Rotations, reflections, and translations are all familiar isometries. *Translation* by the vector $\mathbf{v}$ is the function $\tau: \mathbb{R}^2 \to \mathbb{R}^2$ defined by $\tau(\mathbf{x}) = \mathbf{v} + \mathbf{x}$ for all $\mathbf{x} \in \mathbb{R}^2$. Since $\tau(\mathbf{0}) = \mathbf{v}$, a translation is completely determined if we know where it sends the origin.

We begin by showing that a general element of E_2 is *either a rotation about the origin followed by a translation, or a reflection in a line through the origin*

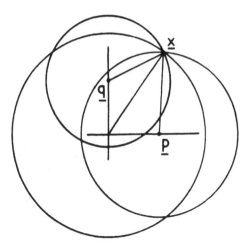

Figure 24.1

followed by a translation. To this end, suppose $g \in E_2$ sends $\mathbf{0}$ to the point $\mathbf{v}$. Let τ denote translation by $\mathbf{v}$. Then the composite isometry $f = \tau^{-1}g$ *fixes the origin.* We claim that f is either a rotation about $\mathbf{0}$ or a reflection in a line through $\mathbf{0}$. Take $\mathbf{p} = (1,0)$ and $\mathbf{q} = (0,1)$ as reference points and agree to abbreviate $f(\mathbf{x})$ to $\mathbf{x}'$ throughout. Each point $\mathbf{x}$ of $\mathbb{R}^2$ is completely determined by the three measurements

$$\|\mathbf{x}\|, \qquad \|\mathbf{x} - \mathbf{p}\|, \qquad \|\mathbf{x} - \mathbf{q}\|$$

because three circles whose centres are not collinear intersect in at most one point (Figure 24.1). For the same reason $\mathbf{x}'$ is determined by its distance from $\mathbf{0}$, $\mathbf{p}'$ and $\mathbf{q}'$. But

$$\|\mathbf{x}'\| = \|\mathbf{x}\|, \qquad \|\mathbf{x}' - \mathbf{p}'\| = \|\mathbf{x} - \mathbf{p}\| \qquad \text{and} \qquad \|\mathbf{x}' - \mathbf{q}'\| = \|\mathbf{x} - \mathbf{q}\|.$$

Therefore, once we know the positions of $\mathbf{p}'$ and $\mathbf{q}'$ we know the effect of f on *every* point of the plane. There are two possibilities as shown in Figure 24.2. Since

$$\|\mathbf{p}'\| = \|\mathbf{p}\| = 1, \qquad \|\mathbf{q}'\| = \|\mathbf{q}\| = 1 \qquad \text{and} \qquad \|\mathbf{p}' - \mathbf{q}'\| = \|\mathbf{p} - \mathbf{q}\| = \sqrt{2},$$

the angle $\mathbf{p}'\mathbf{0}\mathbf{q}'$ is a *right angle.* So if anticlockwise rotation through θ takes $\mathbf{p}$ to $\mathbf{p}'$, then $\mathbf{q}$ rotates either to $\mathbf{q}'$ or to $-\mathbf{q}'$. In the first case f is anticlockwise rotation through θ about the origin, and in the second case f is reflection in the line through the origin which subtends an angle of $\theta/2$ with the positive x-axis. As g consists of f followed by τ our argument is complete.

The *translations* make up a subgroup T of E_2. This is easily checked using (5.1). If $\tau_1, \tau_2 \in T$ are defined by $\tau_1(\mathbf{x}) = \mathbf{u} + \mathbf{x}$, $\tau_2(\mathbf{x}) = \mathbf{v} + \mathbf{x}$ for all $\mathbf{x} \in \mathbb{R}^2$, then

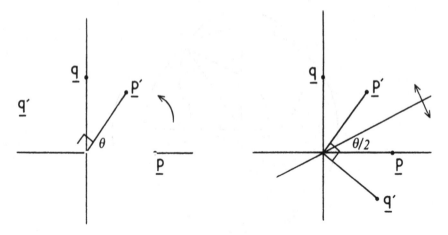

Figure 24.2

$$\tau_1 \tau_2^{-1}(\mathbf{x}) = \tau_1((-\mathbf{v}) + \mathbf{x})$$

$$= \mathbf{u} + ((-\mathbf{v}) + \mathbf{x})$$

$$= (\mathbf{u} - \mathbf{v}) + \mathbf{x}.$$

So $\tau_1 \tau_2^{-1}$ is translation by $\mathbf{u} - \mathbf{v}$ and therefore belongs to T. Let O denote the subgroup of E_2 which consists of the *orthogonal transformations*. In other words the elements of O are rotations about the origin and reflections in lines through the origin. The discussion in the previous paragraph shows that $E_2 = TO$.

The intersection of T and O is just the identity transformation because every non-trivial translation moves the origin, whereas every element of O keeps the origin fixed. The usual argument now shows that each isometry can be written in *only one* way as an orthogonal transformation followed by a translation. For if $g = \tau f = \tau' f'$ where $\tau, \tau' \in T$ and $f, f' \in O$, then $(\tau')^{-1}\tau = f'f^{-1}$ lies in $T \cap O$ and hence $\tau = \tau', f = f'$. If $g = \tau f$ and if f is a rotation, then g is called a **direct isometry**. In the other case, when f is a reflection, g is said to be an **opposite isometry**.

Suppose $f \in O$, $\tau \in T$ and $\tau(0) = \mathbf{v}$. Then for each $\mathbf{x} \in \mathbb{R}^2$ we have

$$f\tau f^{-1}(\mathbf{x}) = f(\mathbf{v} + f^{-1}(\mathbf{x}))$$

$$= f(\mathbf{v}) + f(f^{-1}(\mathbf{x})) \qquad \text{because } f \text{ is linear}$$

$$= f(\mathbf{v}) + \mathbf{x}.$$

Therefore the conjugate $f\tau f^{-1}$ is translation by the vector $f(\mathbf{v})$. Since the elements of T and O together generate E_2, we see (using (15.2)) that T is a *normal* subgroup of E_2.

We can now understand the product structure of our group in terms of the decomposition $E_2 = TO$. If $g = \tau f, h = \tau_1 f_1$ where $\tau, \tau_1 \in T$ and $f, f_1 \in O$, then

$$gh = \tau f \tau_1 f_1 = (\tau f \tau_1 f^{-1})(f f_1)$$

expresses gh as an orthogonal transformation followed by a translation. Put another way the correspondence

$$g \to (\tau, f)$$

is an isomorphism between E_2 and the *semidirect product* $T \times_\varphi O$ where $\varphi: O \to \text{Aut}(T)$ is given by conjugation.

Specific calculations are best carried out using rather different notation. Suppose $g = \tau f$ where $\tau \in T$ and $f \in O$. If $\mathbf{v} = \tau(\mathbf{0})$, and if M is the orthogonal matrix which represents f in the standard basis for $\mathbb{R}^2$, then

$$g(\mathbf{x}) = \mathbf{v} + f_M(\mathbf{x}) = \mathbf{v} + \mathbf{x}M^t \qquad (*)$$

for all $\mathbf{x} \in \mathbb{R}^2$. Conversely, given $\mathbf{v} \in \mathbb{R}^2$ and $M \in O_2$, the formula $(*)$ determines an isometry of the plane. We may therefore think of each isometry as an *ordered pair* $(\mathbf{v}, M)$ in which $\mathbf{v} \in \mathbb{R}^2$ and $M \in O_2$, with multiplication given by

$$(\mathbf{v}, M)(\mathbf{v}_1, M_1) = (\mathbf{v} + f_M(\mathbf{v}_1), MM_1).$$

If we are pressed to be very precise we explain that we have identified E_2 with the semidirect product $\mathbb{R}^2 \times_\psi O_2$, the homomorphism $\psi: O_2 \to \text{Aut}(\mathbb{R}^2)$ being the usual action of O_2 on $\mathbb{R}^2$. Notice that $(\mathbf{v}, M)$ is a direct isometry when $\det M = +1$ and an opposite isometry when $\det M = -1$.

The "simplest" isometries are easily described as ordered pairs. Let

$$A = \begin{bmatrix} \cos \theta & -\sin \theta \\ \sin \theta & \cos \theta \end{bmatrix}, \qquad B = \begin{bmatrix} \cos \varphi & \sin \varphi \\ \sin \varphi & -\cos \varphi \end{bmatrix}$$

and let l, m be the lines shown in Figure 24.3.

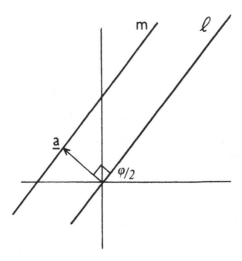

Figure 24.3

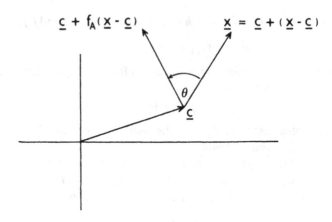

Figure 24.4

(a) *Translation* by the vector $\mathbf{v}$ becomes $(\mathbf{v}, I)$ where I is the 2×2 identity matrix.
(b) *Rotation* anticlockwise through θ about the origin is $(\mathbf{0}, A)$.
(c) *Reflection* in the line l is $(\mathbf{0}, B)$.
(d) *Rotation* anticlockwise through θ about the point $\mathbf{c}$ is $(\mathbf{c} - f_A(\mathbf{c}), A)$.
(e) *Reflection* in the line m is $(2\mathbf{a}, B)$. Notice that $f_B(\mathbf{a}) = -\mathbf{a}$.

Both (d) and (e) require some explanation. Rotation about $\mathbf{c}$ sends each vector $\mathbf{x}$ to $\mathbf{c} + f_A(\mathbf{x} - \mathbf{c})$ as shown in Figure 24.4, and

$$\mathbf{c} + f_A(\mathbf{x} - \mathbf{c}) = \mathbf{c} + f_A(\mathbf{x}) - f_A(\mathbf{c})$$
$$= (\mathbf{c} - f_A(\mathbf{c})) + f_A(\mathbf{x})$$

as required. To realise reflection in m we can first translate by $-\mathbf{a}$ so that m goes onto l, then reflect in l and finally translate l back to m. So $\mathbf{x}$ goes to $\mathbf{x} - \mathbf{a}$, then to $f_B(\mathbf{x} - \mathbf{a})$ and finally to

$$\mathbf{a} + f_B(\mathbf{x} - \mathbf{a}) = \mathbf{a} + f_B(\mathbf{x}) - f_B(\mathbf{a})$$
$$= \mathbf{a} + f_B(\mathbf{x}) + \mathbf{a}$$
$$= 2\mathbf{a} + f_B(\mathbf{x}).$$

A reflection in a line followed by a translation parallel to the same line is called a **glide reflection**. If we take m as the line then each glide reflection along m has the form

$$(2\mathbf{a} + \mathbf{b}, B)$$

where $f_B(\mathbf{b}) = \mathbf{b}$, and $\mathbf{b} \neq \mathbf{0}$.

(24.1) Theorem. *Every direct isometry is a translation or a rotation. Every opposite isometry is a reflection or a glide reflection.*

Proof. A direct isometry is represented by an ordered pair (v, A) where $0 \leqslant \theta < 2\pi$. When $\theta = 0$ we have the *translation* (v, I). Otherwise

$$\det(I - A) = \det \begin{bmatrix} 1 - \cos\theta & \sin\theta \\ -\sin\theta & 1 - \cos\theta \end{bmatrix} = 2 - 2\cos\theta$$

is positive. Therefore, $I - A$ is invertible and the equation

$$c - f_A(c) = f_{I-A}(c) = v$$

has a unique solution for c. The given isometry is the *rotation* $(c - f_A(c), A)$ about this point c.

Each opposite isometry is represented by an ordered pair (v, B) where $0 \leqslant \varphi < 2\pi$. If $f_B(v) = -v$, we have *reflection* in the line m for which $a = v/2$. When $f_B(v) \neq -v$ we set $w = v - f_B(v)$ and observe that

$$f_B(w) = f_B(v - f_B(v))$$
$$= f_B(v) - f_B^2(v)$$
$$= f_B(v) - v = -w.$$

Resolving v along w as in Figure 24.5 gives the vector $(v \cdot w/\|w\|^2)w$ and our isometry is the *glide reflection* $(2a + b, B)$, where $2a = (v \cdot w/\|w\|^2)w$, $b = v - 2a$.

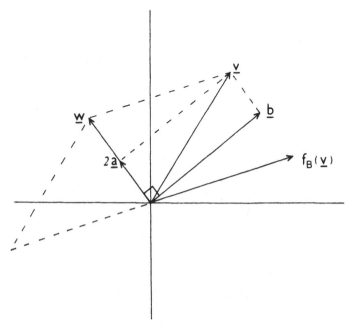

Figure 24.5

EXAMPLE (i). The function $g\colon \mathbb{R}^2 \to \mathbb{R}^2$ given by

$$g(x, y) = \left(1 + \frac{1}{\sqrt{2}}(x - y), 1 - \sqrt{2} + \frac{1}{\sqrt{2}}(x + y)\right)$$

is the isometry which corresponds to the ordered pair $(\mathbf{v}, M)$, where $\mathbf{v} = (1, 1 - \sqrt{2})$ and

$$M = \begin{bmatrix} \dfrac{1}{\sqrt{2}} & -\dfrac{1}{\sqrt{2}} \\ \dfrac{1}{\sqrt{2}} & \dfrac{1}{\sqrt{2}} \end{bmatrix}.$$

Since $\cos(\pi/4) = \sin(\pi/4) = 1/\sqrt{2}$, this is an anticlockwise *rotation* through $\pi/4$. Its centre $\mathbf{c}$ is given by

$$\mathbf{c} - f_M(\mathbf{c}) = \mathbf{c} - \mathbf{c}M^t = \mathbf{v},$$

so that

$$\mathbf{c} = \mathbf{v}(I - M^t)^{-1}$$

$$= (1, 1 - \sqrt{2}) \begin{bmatrix} 1 - \dfrac{1}{\sqrt{2}} & -\dfrac{1}{\sqrt{2}} \\ \dfrac{1}{\sqrt{2}} & 1 - \dfrac{1}{\sqrt{2}} \end{bmatrix}^{-1}$$

$$= (1, 1).$$

Therefore, g is anticlockwise rotation through $\pi/4$ about the point $(1, 1)$.

EXAMPLE (ii). If $h\colon \mathbb{R}^2 \to \mathbb{R}^2$ is defined by

$$h(x, y) = (-\tfrac{1}{2}(x + \sqrt{3}y), 4 + \tfrac{1}{2}(y - \sqrt{3}x)),$$

then h is the isometry $(\mathbf{v}, M)$ where $\mathbf{v} = (0, 4)$ and

$$M = \begin{bmatrix} -\dfrac{1}{2} & -\dfrac{\sqrt{3}}{2} \\ -\dfrac{\sqrt{3}}{2} & \dfrac{1}{2} \end{bmatrix}.$$

This matrix M represents a reflection and

$$f_M(\mathbf{v}) = \mathbf{v}M^t = (-2\sqrt{3}, 2).$$

Since $f_M(\mathbf{v})$ is not equal to $-\mathbf{v}$, the given isometry is a glide reflection. With the notation established earlier, we have

$$w = v - f_M(v) = (2\sqrt{3}, 2),$$

$$v \cdot w = 8, \qquad \|w\|^2 = 16,$$

$$2a = (v \cdot w / \|w\|^2)w = (\sqrt{3}, 1),$$

and

$$b = v - 2a = (-\sqrt{3}, 3).$$

The line of our glide goes through **a** and is parallel to **b**, so its equation is

$$y - \frac{1}{2} = -\sqrt{3}\left(x - \frac{\sqrt{3}}{2}\right) \qquad \text{or} \qquad \sqrt{3}x + y = 2.$$

To apply h we first reflect in this line, then translate by $b = (-\sqrt{3}, 3)$.

EXERCISES

24.1. Write each of the following rotations as an ordered pair (v, M) where $v \in \mathbb{R}^2$ and $M \in O_2$.

(a) Anticlockwise rotation through $\frac{\pi}{6}$ about the point $(-1, -1)$.
(b) Clockwise rotation through $\frac{\pi}{3}$ about the point $(1, 2)$.

24.2. Express reflection in the line $x + y + 3 = 0$, and reflection in the line $\sqrt{3}y - x = 4$, as ordered pairs.

24.3. A rotation through π is usually called a *half-turn*. Prove that the product of two half-turns is always a translation.

24.4. Show that every isometry can be expressed as the product of either two or three reflections.

24.5. Prove that every opposite isometry can be decomposed as a reflection followed by a half-turn.

24.6. If h is a glide reflection with axis m, and if g is an isometry, show that ghg^{-1} is a glide along the line $g(m)$.

24.7. Show that reflection in the line m followed by reflection in m' is a translation when m is parallel to m' and a rotation otherwise.

24.8. Prove that

$$f(x, y) = \left(3 - \sqrt{3} - \frac{1}{2}x - \frac{\sqrt{3}}{2}y, -3 - \sqrt{3} + \frac{\sqrt{3}}{2}x - \frac{1}{2}y\right)$$

is a rotation. Find its centre and the angle through which points are rotated.

24.9. Show that

$$g(x, y) = \left(\frac{3}{5}x + \frac{4}{5}y - 14, \frac{4}{5}x - \frac{3}{5}y + 3\right)$$

represents a glide reflection. Find the axis of the glide and the amount by which points are translated along this axis.

24.10. Prove that

$$h(x, y) = \left(6 - \frac{5}{13}x - \frac{12}{13}y, \, 4 - \frac{12}{13}x + \frac{5}{13}y \right)$$

is a reflection and find its mirror.

CHAPTER 25

Lattices and Point Groups

Figure 25.1 shows a repeating pattern of hexagons which, if continued indefinitely, *fills out the whole plane*. The pattern has a certain amount of symmetry. For example, if we apply either of the translations τ_1, τ_2 or reflect in the x-axis, or rotate anticlockwise through $\pi/3$ about the origin, then hexagons go to hexagons and the pattern is preserved. By shading in part of each hexagon, as in Figure 25.2, we produce a new design which is "less symmetrical" because the rotational symmetry has been destroyed. As usual, the symmetry is measured by a group, in this case the appropriate subgroup of E_2 whose elements are the isometries of the plane which send a given pattern to itself. We shall classify the groups which can arise in this way as symmetry groups of two dimensional repeating patterns or, as we shall call them, *wallpaper patterns*. If you find hexagons rather dull for a wallpaper, then try the designs shown in Figure 25.3. Both exhibit precisely the same symmetry as the pattern of (unshaded) hexagons.

We take on board all the ideas and notation of the previous chapter. In particular we denote each isometry of the plane by an ordered pair $(\mathbf{v}, M)$ where $\mathbf{v} \in \mathbb{R}^2$ and $M \in O_2$. Recall that if $g = (\mathbf{v}, M)$, then

$$g(\mathbf{x}) = \mathbf{v} + f_M(\mathbf{x}) = \mathbf{v} + \mathbf{x}M^t$$

for all $\mathbf{x} \in \mathbb{R}^2$. Define $\pi \colon E_2 \to O_2$ by $\pi(\mathbf{v}, M) = M$. Then π is a homomorphism because

$$\pi((\mathbf{v}, M)(\mathbf{v}_1, M_1)) = \pi(\mathbf{v} + f_M(\mathbf{v}_1), MM_1)$$

$$= MM_1$$

$$= \pi(\mathbf{v}, M)\pi(\mathbf{v}_1, M_1)$$

and its kernel consists of the isometries $(\mathbf{v}, I)$, $\mathbf{v} \in \mathbb{R}^2$, in other words of the

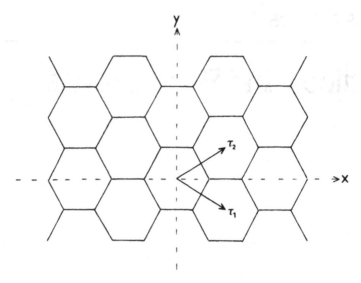

Figure 25.1

translations. If G is a subgroup of E_2, we write H for $G \cap T$ and J for $\pi(G)$, calling H the **translation subgroup** of G and J the **point group** of G. For the symmetry group of the unshaded pattern of hexagons H is generated by the translations τ_1, τ_2 and J is the copy of D_6 in O_2 determined by the two matrices

$$\begin{bmatrix} \cos\dfrac{\pi}{3} & -\sin\dfrac{\pi}{3} \\ \sin\dfrac{\pi}{3} & \cos\dfrac{\pi}{3} \end{bmatrix}, \quad \begin{bmatrix} 1 & 0 \\ 0 & -1 \end{bmatrix}.$$

With the shading added H remains the same but J is reduced to the subgroup

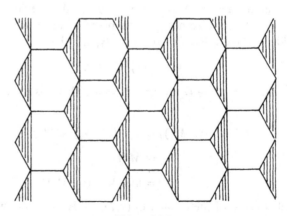

Figure 25.2

of O_2 of order 2 generated by

$$\begin{bmatrix} 1 & 0 \\ 0 & -1 \end{bmatrix}.$$

The restriction of π to G is a surjective homomorphism from G to J whose kernel is H. Therefore, J is isomorphic to the quotient group G/H by the First Isomorphism Theorem.

Here is a precise description of the groups which we shall classify.

*A subgroup of E_2 is a **wallpaper group** if its translation subgroup is generated by two independent translations and its point group is finite.*

The classification will be carried out in Chapter 26. First we need to build up some information about the translation subgroup and the point group of a wallpaper group.

From now on G will denote a wallpaper group with translation subgroup H and point group J. Let L be the orbit of the origin under the action of H on $\mathbb{R}^2$. The set L certainly contains two independent vectors because H is generated by two independent translations. Select a non-zero vector $\mathbf{a}$ of minimum length in L, then choose a second vector $\mathbf{b}$ from L which is skew to $\mathbf{a}$ and whose length is as small as possible.

(25.1) Theorem. *The set L is the **lattice** spanned by $\mathbf{a}$ and $\mathbf{b}$. That is to say, L consists of all linear combinations $m\mathbf{a} + n\mathbf{b}$ where $m, n \in \mathbb{Z}$.*

Proof. The correspondence $(\mathbf{v}, I) \to \mathbf{v}$ is an isomorphism between T and the additive group $\mathbb{R}^2$ which sends H to L. Therefore L is a *subgroup* of $\mathbb{R}^2$ and every point $m\mathbf{a} + n\mathbf{b}$ of the lattice spanned by $\mathbf{a}$ and $\mathbf{b}$ belongs to L. Using the points of this lattice we can divide up the plane into parallelograms as illustrated in Figure 25.4. If $\mathbf{x}$ belongs to L yet is not in the lattice, choose a parallelogram which contains $\mathbf{x}$ and a corner $\mathbf{c}$ of this parallelogram which is as close as possible to $\mathbf{x}$. Then the vector $\mathbf{x} - \mathbf{c}$ is not the zero vector, is not equal to $\mathbf{a}$ or to $\mathbf{b}$, and its length is less than $\|\mathbf{b}\|$. But $\mathbf{x} - \mathbf{c}$ belongs to L because $\mathbf{x}$ and $\mathbf{c}$ both lie in L. We cannot have $\|\mathbf{x} - \mathbf{c}\| < \|\mathbf{a}\|$ since $\mathbf{a}$ is supposed to be of *minimum* length in L. On the other hand, if $\|\mathbf{a}\| \leqslant \|\mathbf{x} - \mathbf{c}\| < \|\mathbf{b}\|$, then $\mathbf{x} - \mathbf{c}$ must be skew to $\mathbf{a}$ and contradicts our choice of $\mathbf{b}$. Therefore, no such point $\mathbf{x}$ can exist and L is the lattice spanned by $\mathbf{a}$ and $\mathbf{b}$. $\square$

We shall classify lattices into *five* different types according to the shape of the basic parallelogram determined by the vectors $\mathbf{a}$ and $\mathbf{b}$. From properties of the lattice of G and the point group of G we plan to build up information

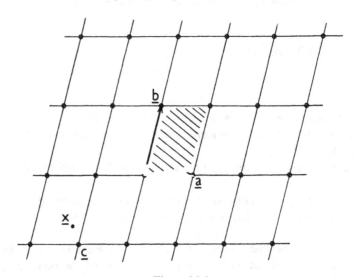

Figure 25.4

about G itself. Replace $\mathbf{b}$ by $-\mathbf{b}$ if necessary to ensure that

$$\|\mathbf{a} - \mathbf{b}\| \leqslant \|\mathbf{a} + \mathbf{b}\|.$$

With this assumption the different lattices are defined as follows.

(a) *Oblique* $\|\mathbf{a}\| < \|\mathbf{b}\| < \|\mathbf{a} - \mathbf{b}\| < \|\mathbf{a} + \mathbf{b}\|$
(b) *Rectangular* $\|\mathbf{a}\| < \|\mathbf{b}\| < \|\mathbf{a} - \mathbf{b}\| = \|\mathbf{a} + \mathbf{b}\|$
(c) *Centred Rectangular* $\|\mathbf{a}\| < \|\mathbf{b}\| = \|\mathbf{a} - \mathbf{b}\| < \|\mathbf{a} + \mathbf{b}\|$
(d) *Square* $\|\mathbf{a}\| = \|\mathbf{b}\| < \|\mathbf{a} - \mathbf{b}\| = \|\mathbf{a} + \mathbf{b}\|$
(e) *Hexagonal* $\|\mathbf{a}\| = \|\mathbf{b}\| = \|\mathbf{a} - \mathbf{b}\| < \|\mathbf{a} + \mathbf{b}\|$.

A glance at Figure 25.5 shows why we use these names. At first sight we appear to have forgotten the possibility

$$\|\mathbf{a}\| = \|\mathbf{b}\| < \|\mathbf{a} - \mathbf{b}\| < \|\mathbf{a} + \mathbf{b}\|.$$

Here the basic parallelogram is a rhombus. Now the diagonals of a rhombus bisect one another at right angles, so we have a centred rectangular structure whose rectangles are based on the vectors $\mathbf{a} - \mathbf{b}$ and $\mathbf{a} + \mathbf{b}$ (Fig. 25.6).

(Wallpaper groups occur in the literature under a variety of aliases, the most common being "plane crystallographic group". If we imagine a similar scenario in *three dimensions*, the corresponding lattice is spanned by three independent vectors and gives a configuration of points which models the internal atomic structure found in *crystals*.)

The point group J is, by its very definition, a subgroup of O_2. However, *there may be no copy of J inside G*. Consider the wallpaper group which is generated by the translation $\tau(x, y) = (x + 1, y)$ and the glide reflection $h(x, y) = (-x, y + 1)$. The line of the glide is the y-axis and is perpendicular to the direction of the translation. A suitable pattern is shown in Figure 25.7. Here the point group is the subgroup

$$J = \left\{ \begin{bmatrix} 1 & 0 \\ 0 & 1 \end{bmatrix}, \begin{bmatrix} -1 & 0 \\ 0 & 1 \end{bmatrix} \right\}$$

of O_2. The group G, however, consists entirely of translations and glide reflections, all of which have infinite order. Therefore, G cannot contain a copy of J. Carrying out the glide reflection h twice gives the translation $(x, y) \rightarrow (x, y + 2)$ and the *lattice* in this example is spanned by the vectors $\mathbf{a} = (1, 0)$, $\mathbf{b} = (0, 2)$. Observe that not all the elements of G send this lattice to itself. None the less the point group does preserve the lattice.

(25.2) Theorem. *The point group J acts on the lattice L.*

Proof. The point group, being a subgroup of O_2, acts on the plane in the usual way. If $M \in J$, and if $\mathbf{x} \in L$, we must show that $f_M(\mathbf{x})$ belongs to L. Suppose $\pi(g) = M$ where $g = (\mathbf{v}, M)$ and let τ denote the translation $(\mathbf{x}, I)$. Since H is the kernel of the homomorphism $\pi: G \rightarrow J$, it is a normal subgroup of G, and therefore $g\tau g^{-1}$ lies in H. But

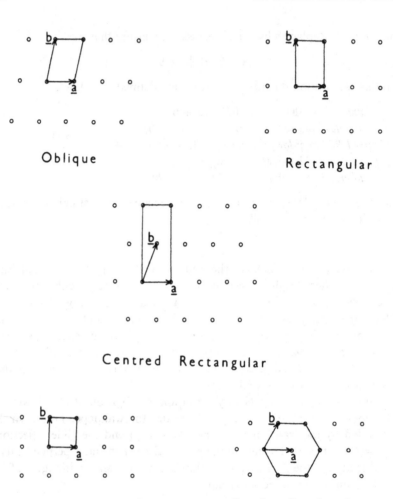

Figure 25.5

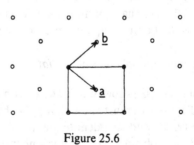

Figure 25.6

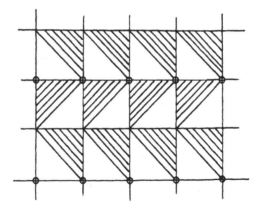

Figure 25.7

$$g\tau g^{-1} = (\mathbf{v}, M)(\mathbf{x}, I)(-f_M^{-1}(\mathbf{v}), M^{-1})$$
$$= (\mathbf{v}, M)(\mathbf{x} - f_M^{-1}(\mathbf{v}), M^{-1})$$
$$= (\mathbf{v} + f_M(\mathbf{x} - f_M^{-1}(\mathbf{v})), MM^{-1})$$
$$= (\mathbf{v} + f_M(\mathbf{x}) - \mathbf{v}, I)$$
$$= (f_M(\mathbf{x}), I),$$

consequently, $f_M(\mathbf{x})$ is a point of the lattice L as required.   □

We recall from (19.1) that finite subgroups of O_2 are either cyclic or dihedral. The next result tells us which of these subgroups can conceivably arise as the point group of a wallpaper group. It is often referred to as the "crystallographic restriction".

(25.3) Theorem. *The order of a rotation in a wallpaper group can only be 2, 3, 4, or 6.*

Proof. Every rotation in a wallpaper group G has finite order because the point group is finite. If we have a rotation of order q, then a suitable power of this rotation is an anticlockwise rotation through $2\pi/q$. Therefore the rotation matrix

$$A = \begin{bmatrix} \cos(\frac{2\pi}{q}) & -\sin(\frac{2\pi}{q}) \\ \sin(\frac{2\pi}{q}) & \cos(\frac{2\pi}{q}) \end{bmatrix}$$

belongs to J. As before, we use $\mathbf{a}$ to denote a non-zero vector of shortest length in the lattice L of G. Now J acts on L, so $f_A(\mathbf{a})$ lies in L. Suppose q is greater than 6. Then $2\pi/q$ is less than 60° and $f_A(\mathbf{a}) - \mathbf{a}$ is a vector in L which is shorter than $\mathbf{a}$ (Fig. 25.8), contradicting our choice of $\mathbf{a}$. If q is equal to five the angle

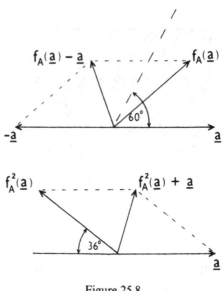

Figure 25.8

between $f_A^2(\mathbf{a})$ and $-\mathbf{a}$ is $36°$. This time $f_A^2(\mathbf{a}) + \mathbf{a}$ lies in L and is shorter than $\mathbf{a}$, and again we have a contradiction.   □

(25.4) Corollary. *The point group of a wallpaper group is generated by a rotation through one of the angles 0, π, $2\pi/3$, $\pi/2$, $\pi/3$ and possibly a reflection.*

Proof. This follows directly from the previous result.   □

(25.5) Theorem. *An isomorphism between wallpaper groups takes translations to translations, rotations to rotations, reflections to reflections and glide reflections to glide reflections.*

Proof. Let $\varphi: G \to G_1$ be an isomorphism between wallpaper groups, and let τ be a translation in G. Translations and glides have infinite order, whereas rotations and reflections are of finite order; therefore, $\varphi(\tau)$ must be either a translation or a glide. Assume $\varphi(\tau)$ is a glide and choose a translation τ_1 from G_1 which does not commute with $\varphi(\tau)$. (Any translation whose direction is *not parallel* to the line of the glide will do.) If $\varphi(g) = \tau_1$, then g has to be a translation or a glide. So g^2 is a translation, and hence commutes with τ, contradicting the fact that $\varphi(g^2) = \tau_1^2$ does not commute with $\varphi(\tau)$. Therefore, translations correspond to translations and glides to glides.

Reflections have order 2, consequently the image of a reflection under an isomorphism is either a reflection or a half-turn. Let $g \in G$ be a reflection whose image $\varphi(g)$ is a half-turn, and choose a translation τ from G in a direction which is *not perpendicular* to the mirror of g. Then τg is a glide. But $\varphi(\tau g) = \varphi(\tau)\varphi(g)$ is the product of a translation and a half-turn, which is another half-turn. Therefore, we have a contradiction and reflections must correspond to reflections. Finally, rotations are now forced to correspond to rotations. □

(25.6) Corollary. *If two wallpaper groups are isomorphic then their point groups are also isomorphic.*

Proof. Let G, G_1 be wallpaper groups with translation subgroups H, H_1 and point groups J, J_1 respectively. If $\varphi: G \to G_1$ is an isomorphism we have $\varphi(H) = H_1$ by (25.5). Therefore, φ induces an isomorphism from G/H to G_1/H_1. The result now follows because J is isomorphic to G/H and J_1 is isomorphic to G_1/H_1. □

EXERCISES

25.1. Which of the following are wallpaper groups?

   (i) The subgroup of E_2 generated by the glides $g(x, y) = (-x, y + 1)$ and $h(x, y) = (-x + 2, y + 1)$.
   (ii) The subgroup of E_2 generated by the translation $\tau(x, y) = (x + 1, y)$ and reflection in the y-axis.
   (iii) The subgroup of E_2 generated by reflection in the x-axis and anticlockwise rotation through $2\pi/3$ about the point $(0, 1)$.
   (iv) The subgroup of E_2 generated by reflection in the x-axis, reflection in the y-axis and reflection in the line $x + y = 1$.

25.2. Sketch the lattices which are spanned by the following pairs of vectors, and state which type of lattice you obtain in each case.

   (i) $\mathbf{a} = (-1, -\sqrt{3})$, $\mathbf{b} = (1, -\sqrt{3})$
   (ii) $\mathbf{a} = (1, 0)$, $\mathbf{b} = (2, -4)$
   (iii) $\mathbf{a} = (-2, 0)$, $\mathbf{b} = (-1, 3)$.

25.3. Let m be a straight line which passes through two points of a lattice L. Prove that m contains infinitely many points of L.

25.4. If L_1 and L_2 are lattices, show that the collection of points of the form $\mathbf{x} + \mathbf{y}$, where $\mathbf{x} \in L_1$ and $\mathbf{y} \in L_2$, is also a lattice.

25.5. What is the result of a half-turn followed by a translation?

25.6. The line fixed by a reflection will be called its **mirror**. Show that two reflections commute if and only if their mirrors either coincide or are perpendicular to one another.

25.7. Prove that a half-turn commutes with a reflection if and only if its centre lies on the mirror of the reflection.

25.8. Show that a translation τ commutes with a reflection g if and only if τ sends the mirror of g to itself.

25.9. Let τ denote translation by $\mathbf{v}$ and g be reflection in the line m. Prove that $g\tau$ is a reflection if $\mathbf{v}$ is perpendicular to m and a glide reflection otherwise.

25.10. Show that tilting a square lattice 45 degrees to the horizontal produces a centred rectangular structure. In how many different ways can we associate a centred rectangular structure with a hexagonal lattice?

25.11. Let L be a lattice spanned by vectors $\mathbf{a}$ and $\mathbf{b}$, and let f be a rotation which preserves L. Taking $\mathbf{a}$ and $\mathbf{b}$ as a basis for $\mathbb{R}^2$ the matrix of f has integer entries, and in particular its trace is an integer. Use the fact that the trace remains invariant under a change of basis to give a second proof of Theorem 25.3.

25.12. Supply an example of a wallpaper group whose point group is isomorphic to Klein's group, and one whose point group is cyclic of order three.

25.13. Let G be a wallpaper group which has a square lattice. What possibilities are there for the point group of G?

25.14. As for the previous exercise, but this time with a hexagonal lattice.

CHAPTER 26

Wallpaper Patterns

There are *seventeen* different wallpaper groups. To see why, we shall examine each of the five possible types of lattice in turn. Given a lattice L we first work out which orthogonal transformations preserve L. Such transformations form a group and, by (25.2), the point group of any wallpaper group which has L as its lattice must be a subgroup of this group. This limitation on the point group is then sufficient to allow us to enumerate the different wallpaper groups with lattice L. An exhaustive analysis of every case would take up too much space. So we concentrate our attention on a small number of examples, and defer the remaining calculations to the exercises. That all the groups we find are genuinely different, in other words that no two are isomorphic, will be shown at the end of the chapter.

Before beginning the classification we add a word or two about notation. Each wallpaper group has a name made up of several (internationally recognised) symbols p, c, m, g and the integers 1, 2, 3, 4, 6. The letter p refers to the lattice and stands for the word *primitive*. When we view a lattice as being made up of primitive cells (copies of the basic parallelogram which do not contain any lattice points in their interiors) we call it a primitive lattice. In one case (the centred rectangular lattice) we take a non-primitive cell together with its centre as the basic building block, and use the letter c to denote the resulting *centred lattice*. The symbol for a reflection is m (for *mirror*) and g denotes a *glide* reflection. Finally, 1 is used for the identity transformation and the numbers 2, 3, 4, 6 indicate rotations of the corresponding order. Rotations of order two are usually called *half turns*.

The seventeen groups are illustrated in Figure 26.1. We show the centres of rotations and the positions of mirrors and glide lines relative to a basic parallelogram. The symbols ○, ▲, □, ● mean that the stabilizer of the corre-

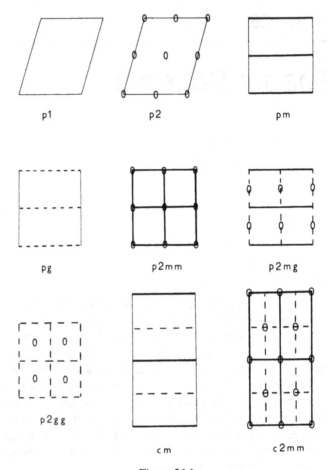

Figure 26.1

sponding point is cyclic of order two, three, four, or six, respectively. Mirrors are drawn as thick lines and glides are indicated by broken lines.

We now proceed with our case-by-case analysis. As usual, G is a wallpaper group with translation subgroup H, point group J, and lattice L. Vectors $\mathbf{a}$ and $\mathbf{b}$ which span the lattice are selected as in Chapter 25. There is no harm in assuming that $\mathbf{a}$ lies along the positive x-axis and that $\mathbf{b}$ is in the first quadrant. Finally, A_θ is the matrix which represents an anticlockwise rotation of θ about the origin, while B_φ represents reflection in the line through the origin which subtends an angle of $\varphi/2$ with the positive x-axis.

Case (a). The lattice of G is *oblique*. Then the only orthogonal transformations which preserve L are the identity and rotation through π about the origin. Therefore, the point group of G is a subgroup of $\{\pm I\}$.

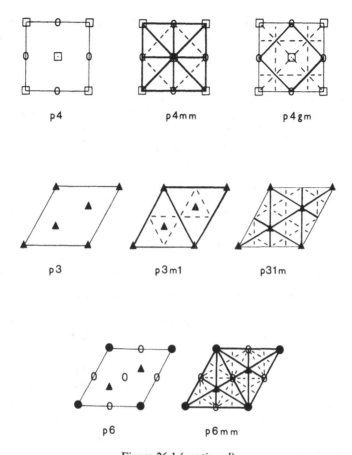

p4   p4mm   p4gm

p3   p3m1   p31m

p6   p6mm

Figure 26.1 (*continued*)

(p1) If J only contains the identity matrix, then G is the simplest of all wallpaper groups; that generated by two independent translations. Its elements have the form $(m\mathbf{a} + n\mathbf{b}, I)$, where $m, n \in \mathbb{Z}$.

(p2) Here J is $\{\pm I\}$. Therefore, G contains a *half turn*, and we may as well take the fixed point of this half turn as origin, so that $(\mathbf{0}, -I)$ belongs to G. The union of the two right cosets H and $H(\mathbf{0}, -I)$ is a subgroup of E_2 which must be our group G. Those elements of G which are not translations lie in $H(\mathbf{0}, -I)$ and have the form

$$(m\mathbf{a} + n\mathbf{b}, I)(\mathbf{0}, -I) = (m\mathbf{a} + n\mathbf{b}, -I)$$

where $m, n \in \mathbb{Z}$. In other words, we have all the half turns about the points $\frac{1}{2}m\mathbf{a} + \frac{1}{2}n\mathbf{b}$.   □

Case **(b).** The lattice of G is *rectangular*. There are now four orthogonal transformations which preserve L; namely, the identity, a half turn about $\mathbf{0}$, reflection in the x-axis, and reflection in the y-axis. Therefore, the point group of G is a subgroup of $\{I, -I, B_0, B_\pi\}$. We look for wallpaper groups *which we have not seen before*, ignoring the possibilities p1, p2 found above.

(pm) J is $\{I, B_0\}$ and G contains a *reflection* in a horizontal mirror.

(pg) Suppose J is $\{I, B_0\}$, yet there are no reflections in G. Then G has to contain a *glide* reflection whose line is horizontal, and we choose a point of this line as origin. Applying a glide reflection twice gives a translation, hence our glide has the form $(\frac{1}{2}k\mathbf{a}, B_0)$ for some integer k. If k is even, then $(-\frac{1}{2}k\mathbf{a}, I)$ is a translation in G, and the reflection

$$(\mathbf{0}, B_0) = (-\tfrac{1}{2}k\mathbf{a}, I)(\tfrac{1}{2}k\mathbf{a}, B_0)$$

belongs to G, contradicting our initial assumption. Therefore, k is *odd* and

$$(\tfrac{1}{2}\mathbf{a}, B_0) = (-\tfrac{1}{2}(k-1)\mathbf{a}, I)(\tfrac{1}{2}k\mathbf{a}, B_0)$$

lies in G. The elements of G which are not translations have the form

$$(m\mathbf{a} + n\mathbf{b}, I)(\tfrac{1}{2}\mathbf{a}, B_0) = ((m + \tfrac{1}{2})\mathbf{a} + n\mathbf{b}, B_0)$$

where $m, n \in \mathbb{Z}$. These are all glides along horizontal lines which either pass through lattice points or lie midway between lattice points. The length of each glide is an odd multiple of $\frac{1}{2}\mathbf{a}$.

Taking $\{I, B_\pi\}$ as point group instead of $\{I, B_0\}$ is tantamount to interchanging the roles of "horizontal" and "vertical" in the preceding discussion, and does not lead to anything new. *From now on we assume that the point group is all of* $\{I, -I, B_0, B_\pi\}$. There are three possibilities according as both, just one, or neither of B_0, B_π can be realised by reflections in G.

(p2mm) In this case G contains a reflection in a horizontal mirror and a reflection in a vertical mirror.

(p2mg) Suppose G contains a reflection in a horizontal mirror but does not contain a reflection in a vertical mirror. Then B_π must be realised in G by a vertical glide reflection. A judicious choice of origin, at the intersection of the horizontal mirror and the vertical glide line, plus the argument used for pg, allow us to assume that $(\mathbf{0}, B_0)$ and $(\frac{1}{2}\mathbf{b}, B_\pi)$ lie in G. The product

$$(\tfrac{1}{2}\mathbf{b}, B_\pi)(\mathbf{0}, B_0) = (\tfrac{1}{2}\mathbf{b}, -I)$$

is the half turn about $\frac{1}{4}\mathbf{b}$. The right cosets

$$H, \quad H(\mathbf{0}, B_0), \quad H(\tfrac{1}{2}\mathbf{b}, B_\pi), \quad H(\tfrac{1}{2}\mathbf{b}, -I)$$

fill out G. In the first of these we have the translations. A typical element of the second has the form

$$(m\mathbf{a} + n\mathbf{b}, I)(\mathbf{0}, B_0) = (m\mathbf{a} + n\mathbf{b}, B_0)$$

where $m, n \in \mathbb{Z}$. When $m = 0$, this isometry is reflection in a horizontal mirror which either passes through lattice points or lies midway between them. If m is not zero, the mirrors change to glide lines and the translation part of the glide is $m\mathbf{a}$. The third coset contains the elements

$$(m\mathbf{a} + (n + \tfrac{1}{2})\mathbf{b}, B_\pi)$$

which are all vertical glides whose lines pass through lattice points or lie midway between them. The translation part of each of these glides is an odd multiple of $\tfrac{1}{2}\mathbf{b}$. Finally, $H(\tfrac{1}{2}\mathbf{b}, -I)$ consists of the half turns centred at the points $\tfrac{1}{2}m\mathbf{a} + \tfrac{1}{2}(n + \tfrac{1}{2})\mathbf{b}$.

Interchanging horizontal and vertical in the preceding discussion leads to a group which is isomorphic to p2mg.

(p2gg) Here there are no reflections in G.   □

Case (c). The lattice of G is **centred rectangular**. The orthogonal transformations which preserve L are the same as in the rectangular case. Therefore, the point group must again be a subgroup of $\{I, -I, B_0, B_\pi\}$. We discover two new groups.

(cm) Suppose J is $\{I, B_0\}$ and that $(\mathbf{v}, B_0\}$ realises B_0 in G. This isometry is either a reflection in a horizontal mirror or a glide along a horizontal line. Choose a point on the mirror or glide line as origin, so that $2\mathbf{v}$ is a multiple of $\mathbf{a}$, and remember that the *vertical* direction is determined by the vector $2\mathbf{b} - \mathbf{a}$.

(i) If $2\mathbf{v} = k\mathbf{a}$ and k is *even*, the *reflection*

$$(\mathbf{0}, B_0) = (-\tfrac{1}{2}k\mathbf{a}, I)(\tfrac{1}{2}k\mathbf{a}, B_0)$$

belongs to G. The elements of G which are not translations have the form

$$(m\mathbf{a} + n\mathbf{b}, B_0) = ((m + \tfrac{1}{2}n)\mathbf{a} + \tfrac{1}{2}n(2\mathbf{b} - \mathbf{a}), B_0)$$

where $m, n \in \mathbb{Z}$. Taking n to be even and $m = -\tfrac{1}{2}n$ produces all the reflections in horizontal mirrors which pass through lattice points. If n is even but $m \neq -\tfrac{1}{2}n$, these mirrors change to glide lines, the translation part of each glide being a multiple of $\mathbf{a}$. Finally, if n is odd, we have glides along lines which lie midway between lattice points. The translation part of each of these glides is an odd multiple of $\tfrac{1}{2}\mathbf{a}$.

(ii) If k is *odd*, then

$$(\tfrac{1}{2}(2\mathbf{b} - \mathbf{a}), B_0) = (-\tfrac{1}{2}(k + 1)\mathbf{a} + \mathbf{b}, I)(\tfrac{1}{2}k\mathbf{a}, B_0)$$

lies in G. This is *again a reflection* and shifting the origin onto its mirror leads back to the previous case.

Substituting $\{I, B_\pi\}$ as point group instead of $\{I, B_0\}$ leads to a group which is isomorphic to cm.

(c2mm) J is $\{I, -I, B_0, B_\pi\}$. The type of calculation carried out above shows that both B_0 and B_π can be realised by reflections in G.   □

Case (d). The lattice of G is *square*. Then the group of orthogonal transformations which preserves L is the dihedral group of order 8 generated by $A_{\frac{\pi}{4}}$ and B_0. The point group J is a subgroup of this group and, to obtain something new, we must include $A_{\frac{\pi}{2}}$ in J. (The other cases are dealt with in Exercise 26.9.)

(p4) Here J is generated by $A_{\frac{\pi}{2}}$.

(p4mm) J is generated by $A_{\frac{\pi}{2}}$ and B_0, and B_0 can be realised by a *reflection* in G.

(p4gm) Suppose J is generated by $A_{\frac{\pi}{2}}$ and B_0, but B_0 *cannot* be realised by a reflection in G. Choose the fixed point of a rotation of order 4 as origin, so that $(0, A_{\frac{\pi}{2}})$ belongs to G, and let $(\lambda\mathbf{a} + \mu\mathbf{b}, B_0)$ realise B_0 in G. Squaring $(\lambda\mathbf{a} + \mu\mathbf{b}, B_0)$ gives $(2\lambda\mathbf{a}, I)$, so 2λ is an integer. If 2λ is even the reflection

$$(\mu\mathbf{b}, B_0) = (-\lambda\mathbf{a}, I)(\lambda\mathbf{a} + \mu\mathbf{b}, B_0)$$

lies in G and we have a contradiction. Therefore, 2λ must be *odd* and

$$(\tfrac{1}{2}\mathbf{a} + \mu\mathbf{b}, B_0) = ((\tfrac{1}{2} - \lambda)\mathbf{a}, I)(\lambda\mathbf{a} + \mu\mathbf{b}, B_0)$$

is an element of G. Also

$$(0, A_{\frac{\pi}{2}})(\tfrac{1}{2}\mathbf{a} + \mu\mathbf{b}, B_0) = (\tfrac{1}{2}\mathbf{b} - \mu\mathbf{a}, B_{\frac{\pi}{2}})$$

and

$$(\tfrac{1}{2}\mathbf{b} - \mu\mathbf{a}, B_{\frac{\pi}{2}})^2 = ((\tfrac{1}{2} - \mu)(\mathbf{a} + \mathbf{b}), I),$$

showing $\tfrac{1}{2} - \mu$ to be an integer. We conclude that the *glide*

$$(\tfrac{1}{2}\mathbf{a} + \tfrac{1}{2}\mathbf{b}, B_0) = ((\tfrac{1}{2} - \mu)\mathbf{b}, I)(\tfrac{1}{2}\mathbf{a} + \mu\mathbf{b}, B_0)$$

belongs to G. The right cosets

$H(0, I),$	$H(0, A_{\frac{\pi}{2}})$
$H(0, -I),$	$H(0, A_{\frac{3\pi}{2}})$
$H(\tfrac{1}{2}\mathbf{a} + \tfrac{1}{2}\mathbf{b}, B_0)$	$H(\tfrac{1}{2}\mathbf{a} + \tfrac{1}{2}\mathbf{b}, B_{\frac{\pi}{2}})$
$H(\tfrac{1}{2}\mathbf{a} + \tfrac{1}{2}\mathbf{b}, B_\pi)$	$H(\tfrac{1}{2}\mathbf{a} + \tfrac{1}{2}\mathbf{b}, B_{\frac{3\pi}{2}})$

fill out G, and it is easy to recognise their elements geometrically. For example, a typical member of $H(\tfrac{1}{2}\mathbf{a} + \tfrac{1}{2}\mathbf{b}, B_{\frac{\pi}{2}})$ has the form

$$((m + \tfrac{1}{2})\mathbf{a} + (n + \tfrac{1}{2})\mathbf{b}, B_{\frac{\pi}{2}})$$
$$= (\tfrac{1}{2}(m + n + 1)(\mathbf{a} + \mathbf{b}) + \tfrac{1}{2}(m - n)(\mathbf{a} - \mathbf{b}), B_{\frac{\pi}{2}})$$

where $m, n \in \mathbb{Z}$. Taking $m + n + 1 = 0$ gives all the reflections in mirrors tilted at 45° to the horizontal which pass midway between lattice points. When $m + n + 1$ is non-zero and $m - n$ is odd, these mirrors change to glide lines. Finally, if $m + n + 1$ is non-zero and $m - n$ is even, we have glides along lines of gradient one which pass through lattice points. The coset $H(0, -I)$ on the other hand contains all the half turns

$(m\mathbf{a} + n\mathbf{b}, -I)$ centred at the points $\frac{1}{2}m\mathbf{a} + \frac{1}{2}n\mathbf{b}$. We leave the reader to work through the remaining cases. $\qquad\qquad\square$

Case (e). The lattice of G is **hexagonal**. Then the point group must be contained in the dihedral group of order 12 generated by $A\frac{\pi}{3}$ and B_0. We are led to new wallpaper groups when J contains rotations of order 3 or 6. (The other cases are dealt with in Exercise 26.10.)

(p3) J is generated by $A\frac{2\pi}{3}$.

(p3m1) J is generated by $A\frac{2\pi}{3}$ and B_0.

(p31m) Suppose J is generated by $A\frac{2\pi}{3}$ and $B\frac{\pi}{3}$. Choose the fixed point of a rotation of order 3 as origin, so that $(\mathbf{0}, A\frac{2\pi}{3})$ belongs to G, and let $(\lambda\mathbf{a} + \mu\mathbf{b}, B\frac{\pi}{3})$ realise $B\frac{\pi}{3}$ in G. Now

$$(\lambda\mathbf{a} + \mu\mathbf{b}, B\tfrac{\pi}{3})^2 = ((\lambda + \mu)(\mathbf{a} + \mathbf{b}), I),$$

so $\lambda + \mu$ is an integer. Also

$$(\mathbf{0}, A\tfrac{2\pi}{3})(\lambda\mathbf{a} + \mu\mathbf{b}, B\tfrac{\pi}{3}) = (\lambda(\mathbf{b} - \mathbf{a}) - \mu\mathbf{a}, B_\pi)$$

and

$$(\lambda(\mathbf{b} - \mathbf{a}) - \mu\mathbf{a}, B_\pi)^2 = (\lambda(2\mathbf{b} - \mathbf{a}), I),$$

showing that λ is an integer. Therefore, both λ and μ are integers and the *reflection*

$$(\mathbf{0}, B\tfrac{\pi}{3}) = (-\lambda\mathbf{a} - \mu\mathbf{b}, I)(\lambda\mathbf{a} + \mu\mathbf{b}, B\tfrac{\pi}{3})$$

belongs to G. The elements of G have the form $(m\mathbf{a} + n\mathbf{b}, M)$ where m, $n \in \mathbb{Z}$ and M is one of the matrices I, $A\frac{2\pi}{3}$, $A\frac{4\pi}{3}$, $B\frac{\pi}{3}$, B_π, $B\frac{5\pi}{3}$. We ask the reader to interpret these elements geometrically. For example

$$(m\mathbf{a} + n\mathbf{b}, B_\pi) = ((m + \tfrac{1}{2}n)\mathbf{a} + \tfrac{1}{2}n(2\mathbf{b} - \mathbf{a}), B_\pi)$$

is a reflection in a vertical mirror when $n = 0$ and a vertical glide otherwise.

(p6) J is generated by $A\frac{\pi}{3}$.

(p6mm) J is generated by $A\frac{\pi}{3}$ and B_0. $\qquad\qquad\square$

Are all these seventeen groups genuinely different? The answer is yes, as we shall see below. By (25.6) we need only concern ourselves with groups whose point groups are isomorphic, so we begin with a summary of the point groups.

G	J	G	J
p1	trivial	p4	$\mathbb{Z}_4$
p2	$\mathbb{Z}_2$	p4mm	D_4
pm	$\mathbb{Z}_2$	p4gm	D_4
pg	$\mathbb{Z}_2$	p3	$\mathbb{Z}_3$
p2mm	$\mathbb{Z}_2 \times \mathbb{Z}_2$	p3ml	D_3
p2mg	$\mathbb{Z}_2 \times \mathbb{Z}_2$	p3lm	D_3
p2gg	$\mathbb{Z}_2 \times \mathbb{Z}_2$	p6	$\mathbb{Z}_6$
cm	$\mathbb{Z}_2$	p6mm	D_6
c2mm	$\mathbb{Z}_2 \times \mathbb{Z}_2$		

Remember that an isomorphism between wallpaper groups sends translations to translations, rotations to rotations, reflections to reflections, and glides to glides.

(26.1) Theorem. *No two of* p2, pm, pg, cm *are isomorphic*

Proof. Among these only p2 contains rotations, so it cannot be isomorphic to any of the others. Of the three remaining groups, pg is the only one which does not contain a reflection; consequently, pg is not isomorphic to pm or cm. Finally, we note that if we take a glide in pm and write it as a reflection followed by a translation, then both the reflection and the translation *belong to pm*. However, cm contains glides whose constituent parts do not lie in cm. For example, consider the glide

$$(\tfrac{1}{2}\mathbf{a} + \tfrac{1}{2}(2\mathbf{b} - \mathbf{a}), B_0) = (\tfrac{1}{2}\mathbf{a}, I)(\tfrac{1}{2}(2\mathbf{b} - \mathbf{a}), B_0).$$

Therefore, pm is not isomorphic to cm. $\square$

(26.2) Theorem. *No two of* p2mm, p2mg, p2gg, c2mm *are isomorphic*

Proof. Among these, p2gg is the only one which does not contain a reflection, so it cannot be isomorphic to any of the others. Of the three remaining groups, only p2mm contains the constituent parts of each of its glides, consequently p2mm is not isomorphic to p2mg or c2mm. Finally, we note that the mirrors of all the reflections in p2mg are *horizontal*, so the product of two reflections is always a translation. But in c2mm there are reflections with horizontal mirrors and reflections with vertical mirrors, and the product of one of each is a half turn. Therefore, p2mg is not isomorphic to c2mm. $\square$

(26.3) Theorem. p4mm *is not isomorphic to* p4gm.

Proof. Each rotation of order 4 in p4mm can be written as the product of two reflections which both belong to p4mm. The corresponding statement is not true for p4gm. For example, $(\mathbf{a}, A_{\frac{\pi}{2}})$ cannot be factorised in p4gm as the product of two reflections (see Exercise 26.4). Therefore, p4mm is not isomorphic to p4gm. $\square$

(26.4) Theorem. p3m1 *is not isomorphic to* p31m.

Proof. In p31m each rotation of order 3 can be written as the product of two reflections, but this is not the case in p3m1. For example, $(\mathbf{a}, A_{\frac{2\pi}{3}})$ cannot be factorised in p3m1 as the product of two reflections (see Exercise 26.5). Therefore, p3m1 is not isomorphic to p31m. $\square$

This completes our classification of wallpaper groups. We have adopted a "hands on" approach, and deliberately so, only by working out *the elements* of these groups do we gain any understanding of their structure. Figure 26.2

Figure 26.2

shows examples of patterns from several different cultures which realise some of the seventeen groups. The three-dimensional case, of interest to crystallographers, is more complicated. There are two hundred and nineteen isomorphism classes of crystallographic groups in three dimensions.

EXERCISES

26.1. Describe the elements of each of the wallpaper groups c2mm, p4mm, p3ml and verify that these groups are represented by the corresponding parts of Figure 26.1.

26.2. Examine each of the patterns of Figure 26.2 in turn and work out its wallpaper group. Compare your answers with those supplied in the figure.

26.3. Find a glide in p2mg whose constituent parts do not lie in p2mg. Do the same for c2mm.

26.4. Prove that $(\mathbf{a}, A\frac{\pi}{2})$ cannot be factorised as the product of two reflections in p4gm.

26.5. Show that $(\mathbf{a}, A\frac{2\pi}{3})$ cannot be factorised as the product of two reflections in p3ml.

26.6. A pattern which repeats in a regular fashion along an infinite strip is called a **frieze**, and the symmetry group of such a pattern is characterised by the requirement that its translation subgroup be infinite cyclic. We may as well imagine the strip to be $\{(x, y) \in \mathbb{R}^2 \,|\, -1 \leqslant y \leqslant 1\}$ and assume that the smallest translation which preserves our pattern is $\tau(x, y) = (x + 1, y)$. Show that (with an appropriate choice of origin) seven possible groups result from frieze patterns; namely, those with the following generators.

   (i) τ.
  (ii) The glide $g(x, y) = (x + \frac{1}{2}, -y)$.
 (iii) τ and the rotation $f(x, y) = (-x, -y)$.
 (iv) τ and the reflection $q(x, y) = (-x, y)$.
  (v) τ and the reflection fq.
 (vi) τ, f, and q.
(vii) g and f.

Invent patterns which realise these groups.

26.7. Notice that groups (i) and (ii) of the previous exercise are both infinite cyclic even though they represent different types of symmetry. Sort the seven "frieze groups" into isomorphism classes. Two frieze groups should be thought of as *equivalent* if they are isomorphic via an isomorphism which sends translations to translations, rotations to rotations, reflections to reflections, and glides to glides. Prove that no two of the groups on our list are equivalent.

26.8. Show that the point group of a frieze group is trivial, cyclic of order 2, or isomorphic to Klein's group.

26.9. Let G be a wallpaper group which has a *square* lattice, so that its point group is a subgroup of

$$\{I, A_{\frac{\pi}{2}}, -I, A_{\frac{3\pi}{2}}, B_0, B_{\frac{\pi}{2}}, B_\pi, B_{\frac{3\pi}{2}}\}.$$

Show that the following table represents all possibilities when the point group does not contain a rotation of order four.

Point Group J	Wallpaper Group G
$\{I\}$	pl
$\{\pm I\}$	p2
$\{I, B_0\}$	pm or pg
$\{I, B_\pi\}$	pm or pg
$\{I, -I, B_0, B_\pi\}$	p2mm, p2mg or p2gg
$\{I, B_{\frac{\pi}{2}}\}$	cm
$\{I, B_{\frac{3\pi}{2}}\}$	cm
$\{I, -I, B_{\frac{\pi}{2}}, B_{\frac{3\pi}{2}}\}$	c2mm

For the last three cases you should tilt the lattice at 45 degrees to the horizontal.

26.10. Let G be a wallpaper group which has a *hexagonal* lattice, so that its point group is a subgroup of the dihedral group generated by $A_{\frac{\pi}{3}}$ and B_0. Show that the following table represents all possibilities when the point group does not contain a rotation of order three.

Point Group J	Wallpaper Group G
$\{I\}$	pl
$\{\pm I\}$	p2
$\{I, B_{\frac{k\pi}{3}}\}, \quad 0 \leqslant k \leqslant 5$	cm
$\{I, -I, B_0, B_\pi\}$	c2mm
$\{I, -I, B_{\frac{\pi}{3}}, B_{\frac{4\pi}{3}}\}$	c2mm
$\{I, -I, B_{\frac{2\pi}{3}}, B_{\frac{5\pi}{3}}\}$	c2mm

You may wish to try Exercise 25.10 first.

CHAPTER 27

Free Groups and Presentations

It is often convenient to be able to describe a group in terms of a set of generators and a set of "relations". For example, the dihedral group D_n is determined by two generators r and s subject to the relations $r^n = e$, $s^2 = e$, $sr = r^{-1}s$, or equivalently $r^n = s^2 = (rs)^2 = e$. We imagine that all the elements of the group can be written as products of powers of r and s, and that the multiplication table is completely determined by the given relations. To make this precise we shall introduce the notion of a free group.

Perhaps the easiest idea to understand is that of a *free set of generators* for a given group. A subset X of a group G is called a free set of generators for G if every $g \in G - \{e\}$ can be expressed in a *unique* way as a product

$$g = x_1^{n_1} x_2^{n_2} \ldots x_k^{n_k} \tag{*}$$

of finite length, where the x_i lie in X, x_i is never equal to x_{i+1}, and each n_i is a non-zero integer. We call the set of generators free because by the uniqueness of $(*)$ there can be no relations between its elements. If G has a free set of generators then it is a *free group*.

Given a nonempty set X we can construct a group which has X as a free set of generators as follows. Define a *word* in the alphabet X to be a finite product

$$x_1^{m_1} x_2^{m_2} \ldots x_s^{m_s}$$

in which each x_i belongs to X and the m_i are all integers, and say that the word is *reduced* if x_i is never equal to x_{i+1} and all the m_i are non-zero. Each word can be simplified to a reduced word by collecting up powers when adjacent elements are equal, and omitting zeroth powers, continuing this process several times if necessary. An example is worth a page of explanation.

Take $X = \{x, y, z\}$ and consider the word

$$w = x^{-3}x^2y^5y^{-5}x^7z^2z^{-2}x^{-1}xzy^2x^{-1}.$$

Then

$$w = x^{-1}y^0x^7z^0x^0zy^2x^{-1}$$
$$= x^{-1}x^7zy^2x^{-1}$$
$$= x^6zy^2x^{-1}$$

which is now reduced. Notice that the elements of X do not commute with one another. It may be possible to carry out the actual process of reduction in more than one way, for example

$$w = (x^{-3}x^2y^5y^{-5}x^7z^2)(z^{-2}x^{-1}xzy^2x^{-1})$$
$$= (x^{-1}y^0x^7z^2)(z^{-2}x^0zy^2x^{-1})$$
$$= (x^{-1}x^7z^2)(z^{-2}zy^2x^{-1})$$
$$= (x^6z^2)(z^{-1}y^2x^{-1})$$
$$= x^6z^2z^{-1}y^2x^{-1}$$
$$= x^6zy^2x^{-1}.$$

However, *the end result is always the same*, as we shall verify in (27.1), and each word w leads to a unique reduced word $\bar{w}$. Reducing the word x_1^0 gives a word with no symbols, which we refer to as the *empty word*. We can multiply words together simply by writing one after the other. If we do this with two reduced words w_1 and w_2, then w_1w_2 may not be reduced because the final symbol of w_1 may be the same as the first symbol of w_2. But w_1w_2 simplifies to a reduced word $\overline{w_1w_2}$, and the set of all reduced words forms a group if we use $\overline{w_1w_2}$ as product. Associativity holds because $(\overline{w_1w_2})w_3$ and $w_1(\overline{w_2w_3})$ result from two different reductions of the word $w_1w_2w_3$, and are therefore both equal to the reduced word $\overline{w_1w_2w_3}$. The identity is the empty word, and the inverse of the reduced word

$$x_1^{n_1}x_2^{n_2}\ldots x_k^{n_k} \text{ is } x_k^{-n_k}\ldots x_2^{-n_2}x_1^{-n_1}$$

which is also reduced. This group of reduced words formed from the alphabet X is called the *free group generated by (the elements of) X* and will be denoted by $F(X)$.

The preceding discussion depends crucially on the following result.

(27.1) Theorem. *Each word can be simplified to only one reduced word.*

Proof. We use an idea reminiscent of the proof of Cayley's theorem. For each $x \in X$ we produce a permutation φ_x of the set of reduced words by the formula

$$\varphi_x(w) = \overline{xw}$$

where w is a *reduced* word. Clearly $\overline{x}w$ is well defined because w is reduced. If

$$u = x_1^{m_1} x_2^{m_2} \ldots x_s^{m_s}$$

is an arbitrary word we have the associated composite permutation

$$\varphi_u = (\varphi_{x_1})^{m_1} (\varphi_{x_2})^{m_2} \ldots (\varphi_{x_s})^{m_s}.$$

Now the permutations of the reduced words form a *group* under composition of permutations and therefore if u, w are words, and if u reduces in some way to w, then $\varphi_u = \varphi_w$. Suppose the same word u can be simplified in two different ways to give the reduced words v and w. Then $\varphi_v = \varphi_u = \varphi_w$. But φ_v sends the empty word to v, and φ_w sends it to w, so we must have $v = w$. □

A free group which is generated by a single element x is infinite cyclic, the only possible reduced words being the powers x^n. When there are two or more generators, $F(X)$ is a non-abelian group in which every element has infinite order. Notice that a bijection φ from X to Y induces an isomorphism between $F(X)$ and $F(Y)$. The reduced word

$$x_1^{n_1} x_2^{n_2} \ldots x_k^{n_k}$$

of $F(X)$ corresponds to the reduced word

$$\varphi(x_1)^{n_1} \varphi(x_2)^{n_2} \ldots \varphi(x_k)^{n_k}$$

of $F(Y)$. Let F_n stand for "the" group which is freely generated by n elements.

(27.2) Theorem. *Abelianising F_n gives $\mathbb{Z}^n$.*

Proof. Let $z_1, \ldots, z_n$ be a free set of generators for F_n. The quotient group $F_n/[F_n, F_n]$ is *abelian* and is generated by the cosets $z_k[F_n, F_n]$, $1 \leqslant k \leqslant n$. By collecting together powers of each generator, every element of this group can be written in precisely one way in the form

$$z_1^{r_1} z_2^{r_2} \ldots z_n^{r_n}[F_n, F_n].$$

The correspondence

$$z_1^{r_1} z_2^{r_2} \ldots z_n^{r_n}[F_n, F_n] \to (r_1, r_2, \ldots, r_n)$$

now provides an isomorphism between $F_n/[F_n, F_n]$ and $\mathbb{Z}^n$. □

(27.3) Theorem. *If F_m is isomorphic to F_n, then $m = n$.*

Proof. For any two groups G, H an isomorphism $\varphi: G \to H$ sends $[G, G]$ to $[H, H]$ because each commutator $xyx^{-1}y^{-1}$ in G corresponds to the commutator $\varphi(x)\varphi(y)\varphi(x)^{-1}\varphi(y)^{-1}$ of H. So φ induces an isomorphism from $G/[G, G]$ to $H/[H, H]$. Therefore, if F_m is isomorphic to F_n we have $\mathbb{Z}^m \cong \mathbb{Z}^n$ and consequently $m = n$. □

Let G be a group and suppose X is a set of generators for G. There is a natural homomorphism π from the free group $F(X)$ to G which sends each reduced word

$$x_1^{n_1} x_2^{n_2} \ldots x_k^{n_k}$$

onto the corresponding product of group elements in G. This homomorphism is surjective because X generates G and, if N denotes its kernel, the First Isomorphism Theorem tells us that $F(X)/N$ is isomorphic to G. *Therefore every group is a quotient of some free group.* It is this result which will allow us to describe groups rigorously in terms of generators and relations. Let R be a collection of words in $F(X)$ which together with all their conjugates generate N. That is to say, N is the smallest normal subgroup of $F(X)$ which contains R. These words determine precisely which words in $F(X)$ become the identity when we pass from $F(X)$ to G or, equivalently, which products of elements of G are the identity in G. We shall call R a set of *defining relations* for G.

EXAMPLE (i). If $G = D_n$ and $X = \{r, s\}$, the words r^n, s^2, $(rs)^2$ form a set of defining relations for G. Let M denote the smallest normal subgroup of $F(X)$ which contains these three words and, as before, use N for the kernel of the homomorphism $\pi: F(X) \to D_n$. We must show that $M = N$. Certainly r^n, s^2, and $(rs)^2$ are all sent to the identity by π, so M is contained in N and by (16.5) we have a surjective homomorphism $F(X)/M \to F(X)/N$ whose kernel is N/M. In particular $|F(X)/M|$ is at least $2n$. Now the cosets rM, sM generate the quotient group $F(X)/M$ and they satisfy

$$(rM)^n = M, \qquad (sM)^2 = M, \qquad (rsM)^2 = M$$

or equivalently

$$(rM)^n = M, \qquad (sM)^2 = M, \qquad srM = r^{n-1}sM \qquad (*)$$

because r^n, s^2, and $(rs)^2$ all belong to M. Using $(*)$ it is easy to show that the cosets

$$M, rM, \ldots, r^{n-1}M, sM, rsM, \ldots, r^{n-1}sM$$

form a *subgroup* of $F(X)/M$. This subgroup contains rM and sM so it has to be the whole group $F(X)/M$. Therefore, $|F(X)/M|$ is at most $2n$. We conclude that $|F(X)/M| = 2n$ and $M = N$, as required.

Take a nonempty set X and let R be a collection of words in the alphabet X. The group determined by the set X of **generators** and the collection R of **defining relations** is defined to be the quotient group $F(X)/N$, where N is the smallest normal subgroup of $F(X)$ which contains R. If G is any group isomorphic to $F(X)/N$ the pair X, R is called a **presentation** for G. In particular, if X is a finite set with elements $x_1, \ldots, x_s$ and if R is a finite collection of words $w_1, \ldots, w_t$ we say that G is **finitely presented** and write

$$G \equiv \{x_1, \ldots, x_s | w_1, \ldots, w_t\}.$$

EXAMPLE (ii). $\mathbb{Z} \equiv \{x|\text{---}\}$.

EXAMPLE (iii). $\mathbb{Z}_n \equiv \{x|x^n\}$.

EXAMPLE (iv). $D_n \equiv \{x, y|x^n, y^2, (xy)^2\}$.

EXAMPLE (v). $Q \equiv \{x, y|x^4, x^2y^{-2}, xyxy^{-1}\}$.

EXAMPLE (vi). For each integer $m \geqslant 2$, the presentation

$$\{x, y|x^{2m}, x^my^{-2}, xyxy^{-1}\}$$

determines a group of order $4m$ called a *dicyclic group*.

EXAMPLE (vii). The same group can of course have many different presentations, for example

$$\mathbb{Z} \equiv \{x|\text{---}\} \equiv \{x, y|y\},$$

$$\mathbb{Z}_6 \equiv \{x|x^6\} \equiv \{x, y|x^3, y^2, xyx^{-1}y^{-1}\}.$$

This second presentation for $\mathbb{Z}_6$ describes it in the form $\mathbb{Z}_3 \times \mathbb{Z}_2$.

EXAMPLE (viii). $\mathbb{Z} \times \mathbb{Z} \equiv \{x, y|xyx^{-1}y^{-1}\}$.

EXAMPLE (ix). The presentation

$$\{x_1, \ldots, x_n|x_ix_jx_i^{-1}x_j^{-1}, 1 \leqslant i < j \leqslant n\}$$

determines a *free abelian group of rank n*.

EXAMPLE (x). Here are presentations for two wallpaper groups.

$$\text{pg} \equiv \{x, y|x^2y^{-2}\}$$

The generators are parallel glide reflections. Using the notation of Section 26 we could take $x = (\frac{1}{2}\mathbf{a}, B_0)$ and $y = (\frac{1}{2}\mathbf{a} + \mathbf{b}, B_0)$.

$$\text{p3ml} \equiv \{x, y|x^2, y^3, (xy^{-1}xy)^3\}.$$

This time the group is generated by a reflection and a rotation of order 3, say $x = (0, B_0)$ and $y = (\mathbf{a}, A\frac{2\pi}{3})$.

We can *characterise* free groups as follows.

(27.4) Theorem. *Let X be a subset of a group G. Then X is a free set of generators for G if and only if given an arbitrary group H, together with a function from X to H, there is a unique extension of this function to a homomorphism from all of G to H.*

Proof. Suppose X is a free set of generators for G. Then each element of $G - \{e\}$ can be written in one and only one way as a reduced word in the alphabet X. Given a group H and a function $f: X \to H$, the *only way* we can extend f to a homomorphism from all of G to H is by sending each reduced word

$$x_1^{n_1} x_2^{n_2} \dots x_k^{n_k}$$

to

$$[f(x_1)]^{n_1} [f(x_2)]^{n_2} \dots [f(x_k)]^{n_k},$$

and the identity of G to that of H.

For the converse we assume that given a group H, together with a function from X to H, we can always find a unique extension of this function to a homomorphism from G to H. As before, let $\pi: F(X) \to G$ be the homomorphism which sends each reduced word to the same product thought of as an element of G. We shall show that π is an *isomorphism*, when X is certainly a free set of generators for G. Taking $H = F(X)$, the inclusion of X into $F(X)$ extends to a homomorphism $\varphi: G \to F(X)$. Clearly, $\varphi\pi$ is the identity from $F(X)$ to $F(X)$ because φ is the identity on X. Also, both $\pi\varphi: G \to G$ and the identity function from G to G are homomorphisms which extend the inclusion of X in G over all of G. By hypothesis there can only be *one* such extension, hence $\pi\varphi$ is the identity from G to G. We conclude that π is an isomorphism.

$\square$

EXERCISES

27.1. Convert the following words in the alphabet $\{x, y, z\}$ into reduced words.

(i) $w_1 = x^{-1} y^3 y^{-1} z^{-2} z z y^{-1} z^{-4} z$
(ii) $w_2 = z^3 y^{-2} x x^{-1} y x^4 z^{-6} z^4$
(iii) $w_3 = z y^5 y^{-2} y^{-3} z^5 x^2 z^{-1} z x^{-3} x z^{-4} x^{-4} y$

27.2. With w_1, w_2, w_3 as above check that $\overline{w_1 w_2} = x^3 z^{-2}$, $\overline{w_2 w_3} = z^3$ and $\overline{w_1 w_2 w_3} = x^{-1} y$.

27.3. Let m and n be positive integers. Prove that there is a homomorphism from F_n onto F_m if and only if m is less than or equal to n.

27.4. Show that F_n contains a normal subgroup of index 2.

27.5. Check in detail that $\{x, y | xyx^{-1} y^{-1}\}$ is a presentation for $\mathbb{Z} \times \mathbb{Z}$.

27.6. Prove that $\{x, y | y^2, (xy)^2\}$ is a presentation for the infinite dihedral group.

27.7. Show that $F_2 \times F_2$ is not a free group. Write down a presentation for $F_2 \times F_2$.

27.8. Find a presentation for each of the seven frieze groups. (These groups were introduced in Exercise 26.6).

27.9. If x, y generate F_2, let

$$X = \{xyx^{-1}, x^2yx^{-2}, x^3yx^{-3}\ldots\}$$

and let H denote the subgroup generated by X. Show that X is a free set of generators for H.

27.10. Let n be a positive integer. Prove that F_2 contains a subgroup which is isomorphic to F_n.

27.11. If G and H are groups we can form words $x_1 x_2 \ldots x_n$ where each x_i lies in the disjoint union $G \cup H$. Call a word reduced this time if x_i and x_{i+1} never belong to the same group and if x_i is never the identity of G or H. Throw in the empty word, and agree to multiply reduced words by juxtaposition, reducing the product as necessary. Show that the result is a group. This group is called the *free product $G * H$* of G and H.

27.12. Let P be a group which contains both G and H as subgroups. Show that P is isomorphic to the free product $G * H$, via an isomorphism which is the identity on both G and H, if and only if given an arbitrary group K, plus a homomorphism from each of G and H to K, there is a unique extension of these homomorphisms to a homomorphism from all of P to K.

27.13. Show that $\mathbb{Z} * \mathbb{Z}$ is isomorphic to F_2.

27.14. Prove that the infinite dihedral group is isomorphic to $\mathbb{Z}_2 * \mathbb{Z}_2$.

Trees and the Nielsen– Schreier Theorem

A *graph* Γ consists of two sets A (directed edges) and V (vertices) together with two functions

$$A \to A, \qquad \alpha \to \bar{\alpha}$$

$$A \to V \times V, \quad \alpha \to (i(\alpha), t(\alpha))$$

which satisfy $\bar{\bar{\alpha}} = \alpha$, $\bar{\alpha} \neq \alpha$ and $i(\bar{\alpha}) = t(\alpha)$ for each α in A. The vertices $i(\alpha)$, $t(\alpha)$ are the *initial* and *terminal* vertices of the directed edge α, and $\bar{\alpha}$ is the *reverse* of α. Henceforth we refer to directed edges simply as edges.

This is very abstract! Luckily we can draw pictures using dots for vertices and arcs for edges as in Figure 28.1(a). In fact our intuition is satisfied by the simplified diagrams of Figure 28.1(b) provided we remember that each physical edge now has to represent a pair of directed edges.

A *path* in Γ joining vertex u to vertex v is an ordered string of edges $\alpha_1 \alpha_2 \ldots \alpha_n$ such that $i(\alpha_1) = u$, $i(\alpha_{k+1}) = t(\alpha_k)$ for $1 \leqslant k \leqslant n - 1$, and $t(\alpha_n) = v$. The special case $\alpha\bar{\alpha}$ where an edge is followed by its reverse is called a *round trip*. A graph is a *tree* if any two distinct vertices may be joined by a path, and if every path which joins a vertex to itself has to contain a round trip. Figure 28.2 illustrates these definitions. If Γ is a tree and if u, v are distinct vertices of Γ, there is *only one* path which joins u to v and which does not contain any round trips. This path is the *geodesic* $\overrightarrow{uv}$ from u to v. Any other path joining u to v can be obtained from this geodesic by successively adding round trips (see Exercise 28.2).

We shall exploit the idea of a group acting on a tree. An *action* of a group G on a graph Γ is an action of G on A and on V such that $g(\bar{\alpha}) = \overline{g(\alpha)}$, $g(i(\alpha)) = i(g(\alpha))$ and $g(\alpha) \neq \bar{\alpha}$ for each g in G and α in A. In other words the elements of G permute the edges and vertices of Γ in a way that is compatible with the structure of Γ as a graph, and no element of G is allowed to reverse

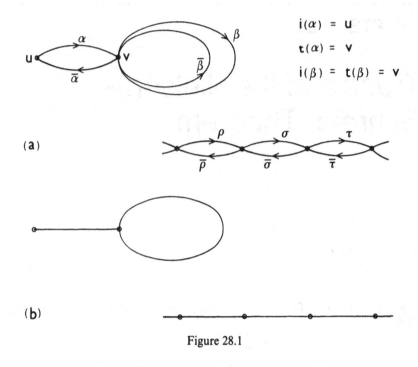

$$i(\alpha) = u$$
$$t(\alpha) = v$$
$$i(\beta) = t(\beta) = v$$

(a)

(b)

Figure 28.1

an edge. The group elements behave correctly on terminal vertices because

$$g(t(\alpha)) = g(i(\bar{\alpha})) = i(g(\bar{\alpha}))$$
$$= i(\overline{g(\alpha)}) = t(g(\alpha))$$

for every edge α. We shall say that G **acts freely** on Γ if the stabilizer of each vertex is just the trivial subgroup $\{e\}$ of G.

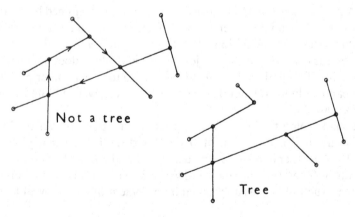

Figure 28.2

EXAMPLE (i). Take a graph whose simplified picture is a letter Y and let G be a cyclic group of order three acting by rotation. Formally we have $A = \{\alpha_r, \bar{\alpha}_r | r = 1, 2, 3\}$ and $V = \{v, v_1, v_2, v_3\}$ with $\bar{\bar{\alpha}}_r = \alpha_r$, $i(\alpha_r) = t(\bar{\alpha}_r) = v$, $t(\alpha_r) = i(\bar{\alpha}_r) = v_r$. If g generates G then the action is determined by $g(\alpha_r) = \alpha_{r+1} \pmod 3$. The stabilizer of v is all of G so in this case G does not act freely.

EXAMPLE (ii). Let A have a pair of edges $\alpha_r, \bar{\alpha}_r$ for each integer r, take V to be Z and set $\bar{\bar{\alpha}}_r = \alpha_r$, $i(\alpha_r) = t(\bar{\alpha}_r) = r$, $t(\alpha_r) = i(\bar{\alpha}_r) = r + 1$. We think of this graph as the real line with the integers marked on as vertices. If G is infinite cyclic with generator g then $g(\alpha_r) = \alpha_{r+1}$ determines a *free* action of G on Γ.

EXAMPLE (iii). Take Γ as in the previous example and let G be the infinite dihedral group with presentation $\{g, h | h^2, (gh)^2\}$. Then $g(\alpha_r) = \alpha_{r+2}$, $h(\alpha_r) = \bar{\alpha}_{-r-1}$ gives an action of G on Γ. In this example G does not act freely because the stabilizer of each vertex is a cyclic group of order 2. To be very explicit the stabilizer of r is $\{e, g^r h\}$.

EXAMPLE (iv). Given a group G and a set X of generators for G we can construct a graph $\Gamma(G, X)$ as follows. For simplicity assume $x^2 \neq e$ if $x \in X$. Let A be the collection of all ordered pairs (g, z) where g comes from G and either z or z^{-1} belongs to X. Take $V = G$ and define

$$\overline{(g, z)} = (gz, z^{-1}),$$

$$i((g, z)) = g, \qquad t((g, z)) = gz.$$

So we have a vertex for each element of G and an edge with initial vertex g_1, and terminal vertex g_2 whenever either $g_1 x = g_2$ or $g_1 x^{-1} = g_2$ for some generator x in X. The action of G on itself by left translation permutes the vertices of $\Gamma(G, X)$ and extends over the edges via $g'((g, z)) = (g'g, z)$ to a *free* action of G on $\Gamma(G, X)$. Extra edges are needed if $x^2 = e$ for some $x \in X$.

Any two vertices g, h of this graph can be joined by a path. Given $g, h \in G$ express $g^{-1}h$ as a product $z_1 z_2 \dots z_k$ of symbols each of which either belongs to X or has its inverse in X. Then $h = gz_1 z_2 \dots z_k$ and the string of edges

$$(g, z_1), (gz_1, z_2), \dots, (gz_1 \dots z_{k-1}, z_k)$$

is a path which starts at g and ends at h.

Here are two special cases. If G is a finite cyclic group of order n, and if X consists of a single generator, we can think of $\Gamma(G, X)$ as a polygon which has n sides with the action of G represented by rotation. If G is a free group generated by $X = \{x, y\}$ the structure of $\Gamma(G, X)$ is illustrated in Figure 28.3. Of course we can only draw the first few stages. We suspect, quite correctly, that this graph is a tree.

(28.1) Theorem. *If X is a free set of generators for G, then $\Gamma(G, X)$ is a tree.*

Proof. Suppose $\alpha_1 \alpha_2 \dots \alpha_n$ is a path which joins the vertex g to itself. If $\alpha_1 = (g, z_1)$, $\alpha_2 = (gz_1, z_2)$, $\dots$, $\alpha_n = (gz_1 \dots z_{n-1}, z_n)$ then $g = gz_1 z_2 \dots z_n$ and

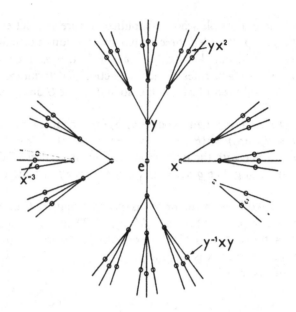

Figure 28.3

therefore $e = z_1 z_2 \ldots z_n$. As X is a free set of generators for G, the empty word is the only reduced word which represents e, so $z_1 z_2 \ldots z_n$ must include a pair of adjacent symbols of the form zz^{-1}. In other words, $\alpha_1 \alpha_2 \ldots \alpha_n$ contains a round trip. Hence, $\Gamma(G, X)$ is a tree.   □

Suppose G acts on the tree Γ. Consider the collection T of all trees inside Γ which contain no more than one edge and one vertex from each orbit. An element Λ of T which is maximal with respect to the partial order given by inclusion will be called a *reference tree*. Maximal means that if Δ also belongs to T, and if $\Lambda \subseteq \Delta$, then $\Lambda = \Delta$. The existence of a maximal element is guaranteed by Zorn's lemma [7]

EXAMPLE (v). Let Γ be the infinite tree which is represented in Figure 28.4. Think of it as a subset of the plane and consider the action of $\mathbb{Z} \times \mathbb{Z}_2 \equiv \{g, h \mid ghg^{-1}h^{-1}, h^2\}$ determined by $g((x, y)) = (x + 3, y)$, $h((x, y)) = (x, -y)$. A reference tree is shown in the diagram. Find all possible reference trees which contain the vertex 0; there are twelve altogether.

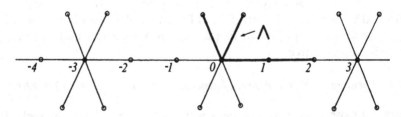

Figure 28.4

(28.2) Theorem. *A reference tree contains precisely one vertex from each orbit.*

Proof. Let Λ be a reference tree for the action of G on the tree Γ. By definition no two vertices of Λ lie in the same orbit. Suppose that Λ does not contain a vertex from each orbit. Select a vertex v from Λ and a vertex z of Γ whose orbit does not meet Λ. Let α be the first edge of $\overrightarrow{zv}$ and set $y = t(\alpha)$. If the orbit of y meets Λ, say $g(y) \in \Lambda$, adding $g(z), g(\alpha), g(\bar{\alpha})$ to Λ produces a larger tree than Λ which contains at most one vertex from each orbit. This contradicts the maximality of Λ. On the other hand, if the orbit of y does not meet Λ we simply replace z by y and repeat our argument until we reach a contradiction. $\qquad\square$

(28.3) Theorem. *If G acts freely on a tree then G is a free group.*

Proof. Suppose G acts freely on the tree Γ and choose a reference tree Λ for the action. Then Λ contains precisely one vertex from each orbit. Let A_* denote the collection of those edges of Γ which, though not in Λ, do have their initial vertices in Λ. Given $\alpha \in A_*$ let z be the vertex of Λ which is in the same orbit as $t(\alpha)$ and choose a group element g_α such that $g_\alpha(z) = t(\alpha)$. There is only one element of G with this property, for if g_α and h_α both send z to $t(\alpha)$ then $h_\alpha^{-1}g_\alpha(z) = z$. But the action is *free*, so $h_\alpha^{-1}g_\alpha$ must be the identity of G and consequently $h_\alpha = g_\alpha$. These edges and group elements come in *pairs*, for the edge $\alpha' = g_\alpha^{-1}(\bar{\alpha})$ also belongs to A_* and its corresponding group element is $g_{\alpha'} = g_\alpha^{-1}$, see Figure 28.5. From each such pair g_α, g_α^{-1} we select one and use X to denote the resulting subset of G. *We shall show that X is a free set of generators for G.*

Choose a vertex v in Λ. Given $g \in G - \{e\}$ let α be the first edge of the geodesic from v to $g(v)$ which is *not* in Λ. Then, of course, α belongs to A_*. Apply g_α^{-1} to the geodesic from $t(\alpha)$ to $g(v)$. The resulting path starts at the vertex $g_\alpha^{-1}(t(\alpha))$ of Λ and ends at $g_\alpha^{-1}g(v)$. Follow this new path to the first edge β where it leaves Λ, then apply g_β^{-1} to the geodesic from $t(\beta)$ to $g_\alpha^{-1}g(v)$. Again, we have a path which begins in Λ, and this time it ends at $g_\beta^{-1}g_\alpha^{-1}g(v)$. Repeat the procedure and notice that the path is shortened at each stage. Therefore, we eventually produce a path which lies *entirely* in Λ. Its end point is say $g_\nu^{-1}\dots g_\beta^{-1}g_\alpha^{-1}g(v)$. As no two vertices of Λ belong to the same orbit, we must have $g_\nu^{-1}\dots g_\beta^{-1}g_\alpha^{-1}g(v) = v$, and since the action is free, this gives $g_\nu^{-1}\dots g_\beta^{-1}g_\alpha^{-1}g = e$ and hence

$$g = g_\alpha g_\beta \dots g_\nu. \tag{$*$}$$

Already we see that the elements of X generate G.

The right hand side of $(*)$ determines a *reduced* word $w(g)$ in symbols from X. Our construction used the geodesic $\overrightarrow{vg(v)}$, but any other path which joins v to $g(v)$ will do equally well and lead to the *same* reduced word. We check first of all that the addition of a single round trip $\sigma\bar{\sigma}$ to the geodesic does not alter $w(g)$. This is easy; just keep track of σ during the above process. Either

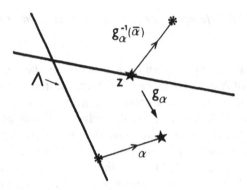

Figure 28.5

σ lands inside Λ at some stage, in which case (∗) is unchanged, or σ goes to an edge τ of A_* when (∗) changes to $g = g_\alpha g_\beta \ldots g_\tau g_\tau^{-1} \ldots g_\nu$. In both cases $w(g)$ is unaltered. Extra round trips may be dealt with in a similar fashion. Now remember that an arbitrary path between two vertices can be obtained from the geodesic by the successive addition of round trips.

To complete our argument we shall show that any decomposition $g = g_\gamma g_\delta \ldots g_\rho$, where $\gamma, \delta, \ldots, \rho$ all lie in A_*, may be realised by a suitably chosen path from v to $g(v)$. We can then be sure that $w(g)$ is the *only* reduced word in the elements of X which represents g. Proceed by induction on the length of the decomposition. Set $g_1 = g_\delta \ldots g_\rho$ and let P be a path from v to $g_1(v)$ which realises this decomposition of g_1. By following $\overrightarrow{vg_\gamma}(v)$ then $g_\gamma(P)$ we produce a path Q from v to $g(v)$. Examine $\overrightarrow{vg_\gamma}(v)$, it begins with $\overrightarrow{vi(\gamma)}$, leaves Λ along γ, then continues via $g_\gamma(\overrightarrow{zv})$ where $z = g_\gamma^{-1}(t(\gamma)) \in \Lambda$. Applying the first step of our process to Q produces g_γ and leaves us with $\overrightarrow{zv}$ (which lies in Λ) followed by P. Therefore Q realises $g_\gamma g_\delta \ldots g_\rho$ as required.   □

(28.4) Nielsen–Schreier Theorem. *Every subgroup of a free group is free.*

Proof. Let F be the free group generated by the set X and let G be a subgroup of F. We know that F acts freely on the tree $\Gamma(F, X)$ and consequently so does G. Hence, G is free by (28.3).   □

<small>EXERCISES</small>

28.1. Draw (simplified) diagrams of the following graphs:

(i) $A = \{\alpha, \bar{\alpha}, \beta, \bar{\beta}, \gamma, \bar{\gamma}, \delta, \bar{\delta}\}$,
$V = \{u, v, w\}$ and
$i(\alpha) = t(\gamma) = t(\delta) = u$,
$t(\alpha) = i(\beta) = i(\delta) = v$,
$t(\beta) = i(\gamma) = w$.

(ii) $A = \{\alpha_r, \bar{\alpha}_r, |1 \leqslant r \leqslant 7\}$
$V = \{v_1, v_2, v_3, v_4\}$ and
$i(\alpha_1) = t(\alpha_4) = i(\alpha_5) = t(\alpha_7) = v_1,$
$t(\alpha_1) = i(\alpha_2) = t(\alpha_6) = v_2,$
$t(\alpha_2) = i(\alpha_3) = t(\alpha_5) = i(\alpha_7) = v_3,$
$t(\alpha_3) = i(\alpha_4) = i(\alpha_6) = v_4.$

28.2. Let u,v be distinct vertices of a tree Γ. Show there is only one path in Γ which joins u to v and which does not contain any round trips. This path is the so called geodesic $\overrightarrow{uv}$. If P is another path from u to v show there are paths $P_1, \ldots, P_k$ in Γ, all of which join u to v, such that P_{r+1} can be obtained from P_r by the addition of a single round trip for $1 \leqslant r \leqslant k - 1$, $P_1 = \overrightarrow{uv}$ and $P_k = P$.

28.3. Suppose we have a group action on a tree Γ. If the group element g fixes two vertices u, v of Γ, prove that g must leave all of the geodesic $\overrightarrow{uv}$ fixed.

28.4. Find reference trees for the group actions described in Examples (i), (ii), and (iii) of this section.

28.5. Work out all possible actions of a cyclic group of order two on each of the graphs described in Exercise 28.1.

28.6. Examine how the proof of Theorem 28.3 works in the situation of Example (ii) of this section, and for the natural action of the free group on two generators on the tree shown in Figure 28.3.

28.7. The elements of $SL_2(\mathbb{Z})$, the group of 2×2 matrices which have integer entries and determinant $+1$, act on the upper half of the complex plane as Möbius transformations. Let α denote the arc of the unit circle which joins $\exp(i\pi/3)$ to $\exp(i\pi/2)$ in the upper half plane and define Γ to be the union of all segments $g(\alpha)$ where $g \in SL_2(\mathbb{Z})$. Draw a picture of Γ and check that it represents a tree. Show that the action of $SL_2(\mathbb{Z})$ on this tree is not a free action.

28.8. Let u,v be vertices of a graph Γ. If P, P' are paths which both join u to v we say that P' is *adjacent* to P provided P' can be obtained from P by the addition or removal of a single round trip. Show that this leads to an equivalence relation on the collection of all paths from u to v in Γ, where P is related to Q provided there are paths $P_1, \ldots, P_k$, all of which join u to v, such that $P_1 = P$, $P_k = Q$ and P_r is adjacent to P_{r+1} for $1 \leqslant r \leqslant k - 1$. How many equivalence classes are there when Γ is a tree?

28.9. As usual Γ is a graph and v is a vertex of Γ. A path P in Γ which joins v to itself will be called a *loop* based at v, and $[P]$ will be used to denote its equivalence class under the equivalence relation introduced in the previous exercise. If $P = \alpha_1\alpha_2 \ldots \alpha_s$ and $Q = \beta_1\beta_2 \ldots \beta_t$ both join u to

v we write PQ for the loop $\alpha_1 \alpha_2 \ldots \alpha_s \beta_1 \beta_2 \ldots \beta_t$. Prove that the collection of all equivalence classes of loops based at v forms a group under the product $[P][Q] = [PQ]$. (You should begin by checking that the product is well defined.) This group is called the **fundamental group** of Γ based at v.

28.10. Adopt the terminology of the previous exercise and assume that any two distinct vertices of Γ may be joined by a path. Let Λ be a maximal tree in Γ. Show that Λ contains every vertex of Γ. Choose an edge from each pair of directed edges which do not lie in Λ, and let X denote the resulting collection of edges. Prove that the fundamental group of Γ based at v is isomorphic to the free group $F(X)$.

Bibliography

[1] Weyl, H., *Symmetry*, Princeton University Press, Princeton, N.J., 1952.
(*A classic essay which should be read by every mathematics teacher and student.*)

[2] Lyndon, R.C., *Groups and Geometry*, Cambridge University Press, Cambridge (England), 1985.
[3] Neumann, P.M., Stoy, G.A., and Thompson, E.C., *Groups and Geometry*, Duplicated notes produced by the Mathematical Institute, Oxford University, 1980.
[4] Rees, E.G., *Notes on Geometry*, Springer-Verlag, Berlin, Heidelberg, 1983.
(*For the role played by groups in geometry.*)

[5] Hill, V.E., *Groups, Representations and Characters*, Hafner Press, New York, 1975.
(*The beginnings of representation theory.*)

[6] Rotman, J.J., *The Theory of Groups: An Introduction*, Allyn and Bacon, Boston, Third edition 1984.
(*An excellent graduate level text.*)

[7] Halmos, P.J., *Naive Set Theory*, Van Nostrand, Princeton N.J., 1960 and Springer-Verlag, New York, 1974.
(*The elements of set theory, including Zorn's lemma. An illustration of how to write mathematics.*)

[8] Jones, O., *The Grammar of Ornament*, Day and Son, London, 1856 and Omega Books, Ware (England), 1986.
(*A wonderful collection of ornamental patterns and designs from different civilisations all over the world.*)

[9] Martin, G.E., *Transformation Geometry*, Springer-Verlag, New York, 1983.
[10] Lockwood, E.H. and Macmillan, R.H., *Geometric Symmetry*, Cambridge University Press, Cambridge (England), 1978.
(*For a wealth of information about frieze, wallpaper and space patterns.*)

Index

Undergraduate Texts in Mathematics

(continued from page ii)

Gamelin: Complex Analysis.

Gordon: Discrete Probability.

Hairer/Wanner: Analysis by Its History. *Readings in Mathematics.*

Halmos: Finite-Dimensional Vector Spaces. Second edition.

Halmos: Naive Set Theory.

Hämmerlin/Hoffmann: Numerical Mathematics. *Readings in Mathematics.*

Harris/Hirst/Mossinghoff: Combinatorics and Graph Theory.

Hartshorne: Geometry: Euclid and Beyond.

Hijab: Introduction to Calculus and Classical Analysis.

Hilton/Holton/Pedersen: Mathematical Reflections: In a Room with Many Mirrors.

Hilton/Holton/Pedersen: Mathematical Vistas: From a Room with Many Windows.

Iooss/Joseph: Elementary Stability and Bifurcation Theory. Second edition.

Isaac: The Pleasures of Probability. *Readings in Mathematics.*

James: Topological and Uniform Spaces.

Jänich: Linear Algebra.

Jänich: Topology.

Jänich: Vector Analysis.

Kemeny/Snell: Finite Markov Chains.

Kinsey: Topology of Surfaces.

Klambauer: Aspects of Calculus.

Lang: A First Course in Calculus. Fifth edition.

Lang: Calculus of Several Variables. Third edition.

Lang: Introduction to Linear Algebra. Second edition.

Lang: Linear Algebra. Third edition.

Lang: Short Calculus: The Original Edition of "A First Course in Calculus."

Lang: Undergraduate Algebra. Second edition.

Lang: Undergraduate Analysis.

Laubenbacher/Pengelley: Mathematical Expeditions.

Lax/Burstein/Lax: Calculus with Applications and Computing. Volume 1.

LeCuyer: College Mathematics with APL.

Lidl/Pilz: Applied Abstract Algebra. Second edition.

Logan: Applied Partial Differential Equations.

Lovász/Pelikán/Vesztergombi: Discrete Mathematics.

Macki-Strauss: Introduction to Optimal Control Theory.

Malitz: Introduction to Mathematical Logic.

Marsden/Weinstein: Calculus I, II, III. Second edition.

Martin: Counting: The Art of Enumerative Combinatorics.

Martin: The Foundations of Geometry and the Non-Euclidean Plane.

Martin: Geometric Constructions.

Martin: Transformation Geometry: An Introduction to Symmetry.

Millman/Parker: Geometry: A Metric Approach with Models. Second edition.

Moschovakis: Notes on Set Theory.

Owen: A First Course in the Mathematical Foundations of Thermodynamics.

Palka: An Introduction to Complex Function Theory.

Pedrick: A First Course in Analysis.

Peressini/Sullivan/Uhl: The Mathematics of Nonlinear Programming.

Prenowitz/Jantosciak: Join Geometries.

Priestley: Calculus: A Liberal Art. Second edition.

Undergraduate Texts in Mathematics